To our readers

Peter Ralston

Looking at things differently

PEOPLE WITH QUIRKY, original minds have long populated Maine's islands, drawn there, perhaps, in the belief they could escape a stifling conformity they experienced on the mainland. An island community may not be the easiest place for the person with an unusual turn of mind to live — it can and will impose strictures of its own — but for the individual willing to hone his or her uniqueness to an island edge, living "enisled" can bring out one's very best. Stimulated by ever-present encounters between land and sea, man and nature, wild and tamed, generations of creative people have accomplished brilliant things on Maine islands. Over the years other qualities — confidence born of learning how to live well with others at close range, knowledge gained from island traditions of craftsmanship and care, new ways of looking at life learned from quiet observation and, simply, the opportunity to think — have served islanders well. Life proceeds at a different pace in an island community than it does in more populous places; for the person accustomed to thinking "outside the box" an island can provide a perfect environment.

Island Journal has long celebrated the arts. This issue carries on that tradition, with greater enthusiasm than ever. We consider what happens when theater enriches an entire island community, young and old. We introduce readers to an artist and children's author who draws inspiration from many places, including the Maine island where he lives. We enter the mind of a world-class novelist, John Fowles, whose connections with islands span a lifetime.

And then, as islanders (and others) of adventurous mind always will, we go exploring: to Labrador, to Newfoundland, to New Brunswick, across the Gulf of Maine and into the newest frontier, cyberspace.

The richness, diversity and sheer adventure of island life have always brought forth creativity. And in a cynical world that so often seems blind and lost, the creative force can show us the way.

The Editors

ISLAND JOURNAL

The Annual Publication of the Island Institute
Volume Sixteen

Dedication	page 4
Log of RAVEN	page 5
A Beneficent Magic	page 6
Drama, the Fox Islands and John Wulp	
By Philip W. Conkling	
Photographs by Bridget Besaw Gorman	
The Mr. Wulp Effect	page 12
By Karen Roberts Jackson	
"Something Different To Look At"	page 14
Street theater comes to Vinalhaven	
By Mike Gorman	
The Dream Keeper	page 19
Ashley Bryan of Little Cranberry Island	
By Susan Hand Shetterly	
Photographs by Bridget Besaw Gorman	
FOLIO	page 24
A Siren Call	
For John Fowles, islands are the most powerful metaphor of all	
Excerpts by John Fowles	
Photographs by Peter Ralston	
Boats and Hoops	page 32
Why do the fastest lobsterboats and the finest basketball players come from the same island?	
By Sandra Dinsmore	
Running Together	page 36
Aquaculture transforms an island community	
By K. J. Vaux	
Photographs by Heather D. Hay	
Hard Times, Good Times and the Seat of the Pants	page 42
Nine Decades on Chebeague	
By Donna Miller Damon	
Evicted	page 48
How the State of Maine destroyed a "different" island community	
By Deborah DuBrule	

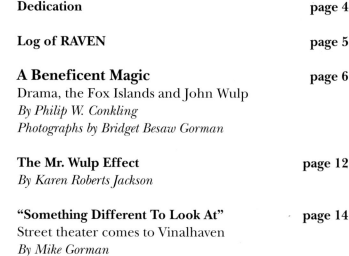

page 54

The Land Called Nunavut   page 54
A Journey Across Baffin Island
By Dylan, Tristan and Hope Jackson

The Defense that Failed   page 60
Had Ann Brown's murderer been tried
30 years later, the verdict might have been different.
By Randy Purinton
Illustration by Siri Beckman

A Fitting Destruction   page 65
The Whaling Station at Hawke Harbour,
Labrador
By John Bockstoce
Photographs by Nicholas Whitman

The Great Lobster Collaboration   page 68
Layer by Layer, Penobscot Bay reveals itself
By Philip W. Conkling

A Certain Obscurity   page 76
Swan's Island informs the poetry
of Donald Junkins

Removing Time and Distance   page 80
As Maine's year-round islands launch
themselves into cyberspace, there's much to gain and
lose — and plenty of uncertainty.
By Brian Willson
Illustration by Robert Shetterly

Frenchboro at War, 1941-1945   page 84
By Dean Lunt

Salt Water Windows   page 88
The stained glass creations of Janet Redfield

Reviews   page 94
Acadia: Visions and Verse
The Publications of the Islesford Historical Society
By Carl Little

Cover: Rockland Breakwater, by Peter Ralston

ISLAND INSTITUTE
Publishers of Island Journal

Serving the Islands and Communities of the Gulf of Maine

ISLAND JOURNAL

PUBLISHER
Philip W. Conkling

EDITOR
David D. Platt

ART DIRECTOR
Peter Ralston

COPY EDITOR
Esme McTighe

GRAPHICS RESEARCH
Bonnie Mowery-Oldham

DESIGN & PRODUCTION
Michael Mahan Graphics
Bath, Maine

PRINTING
The J. S. McCarthy Company
Augusta, Maine

•

ISLAND INSTITUTE

PRESIDENT
Philip W. Conkling

EXECUTIVE VICE PRESIDENT
Peter Ralston

FINANCE AND OPERATIONS
Josee L. Shelley, CPA; Marianne H. Pinkham;
Barbara Nickerson; Charlene French

SCHOOLS AND COMMUNITY SERVICES
Susan Valaitis, Marge Kilkelly

COMMUNITY SCIENCE
Annette S. Naegel

MARINE RESOURCES
Bill MacDonald, Chris Brehme, Carl Wilson, Corrie Roberts, Joe
Mariano, Leslie Fuller

PUBLICATIONS
David D. Platt, Charles G. Oldham, Michael Herbert

MEMBERSHIP AND DEVELOPMENT
John Guarnaccia, Jody A. Herbert, Kathy Estabrook, Steve
McBride, Angie Crossman

RECEPTIONISTS
Kathy Allen, Pat Ritchie

PART TIME
Deborah DuBrule, Lucy Hallowell, Marcia Reisman

BOARD OF TRUSTEES

CHAIRMAN
Horace A. Hildreth, Jr.

TREASURER
John Higgins

SECRETARY
Donna Miller Damon

CLERK
Michael P. Boyd

TRUSTEES
Robert Bass, Jr.
John Bird
Eric Davis
Mary Beth Dolan
William Ginn
Margery Hamlen
Peggy Krementz
David L. Lunt
Nigel MacEwan
George Pew
Peter O. Willauer

page 68

page 88

We dedicate the 1999 Island Journal to
Jamien Elise Morehouse

wife of Philip W. Conkling,
mother of Tim, Sam, James and Micah.

Without her,
there could never have been
an Island Institute.

The annual chronicles of the Island Institute

Peter Ralston

EYE OF THE *RAVEN*
From the logs of the vessels of the Island Institute

PHILIP W. CONKLING

IT WAS A BATTLEFIELD promotion of sorts: taking over as interim captain of RAVEN from Peter Ralston. At the time, Peter was recovering from a series of life-threatening surgeries, a story which has the happiest of all endings in his return to the Institute 13 months later. In the meantime, I would be RAVEN's helmsman and keeper. Understand that RAVEN is a high-bowed black beauty, a real boat even in the eyes of hard-dog lobstermen who know the pros and cons of every lobster hull ever designed in Maine. A Repco hull means something to these men, because they know that these early glass hulls from West Gouldsboro were laid up by hand, with layer on layer of thick fiberglass that have proven to be strong, reliable and durable. From a distance you can see in her lines a sea-kindliness in the way she settles low and steady in the water from the pilothouse aft. And her distinctive downeast lines sweep up in the bow to give you confidence that she can take a nasty head sea if necessary. As you draw toward her or away from her starboard quarter at the beginning or end of a day or an expedition, you know you are privileged to be with her.

The appointment came as the season was turning from raw spring to summer sweetness, time enough before the crazed heat of August to get a feel for RAVEN's helm. Peter carefully went over her systems with me during a two-hour seminar in early May. He presided over the detailed pre-ignition ritual, an intonation with the all-important 3116 Caterpillar Diesel we must revere above all else; he showed me how to trace, check, fill and operate the water system, monitor the bilge and check the pumps. He went over the heating, propane, radio and electrical systems; he described her lines, fenders, ground tackle and a daunting number of other intricate details that had to be kept in mind.

Continued on page 92

JOHN WULP
A Beneficent Magic

PHILIP W. CONKLING

Photographs by Bridget Besaw Gorman

"I have an island in the palm of my right hand. It is quite large and shaped like an almond. To make this island, the fate line splits in two in the middle, and comes together again. I don't know what an island means in palmistry. But it looks to me as if it meant that a quiet respectable fate were suddenly going to explode in the middle of like into something entirely new and strange, and then be folded together again and go on quietly as it began. I treasure that little thing in my hand. I pore over it reminiscently, gratefully. I like to know it is there. It is the lucky coin that saved me. It is the wafer of beneficent magic that made everything all right at last."

— *from* The Little Locksmith, *Katherine Butler Hathaway, 1942*

"I have completely remade my life in this community."

WHERE TO START? At the end or the beginning of John Wulp's story? In a sense it does not much matter, because you could say that his beginnings and his endings are circular— arriving where they started for a stageman, theater director and artist, who washed ashore by accident on a Maine island more than a decade ago. Once there, he discovered the island's community and became the greatest teacher of drama to school children that the state, perhaps, has ever seen. At the North Haven school, John Wulp has achieved unimaginable successes with a drama program in a small and insular place, with its gifts for mimicry. He created not just a powerful new vision within his adopted island community, but recreated himself, one senses, in the process.

Two and a half decades ago, during the 1970s, Wulp was a highly acclaimed, award-winning theater director and producer on and off-Broadway. He had Tony, Obie and Theater Critics awards to his credit as well as his blockbuster production, *Dracula*, which for a time captured and defined the New York theater world, at a time when all that glittered might have been gold. But between these triumphs Wulp traversed the territory of failure and exiled himself to a seemingly desolate place where austerity either drives you under or you reach the deeper routes of self-renewal.

Shipwrecked

THE SETTING: Penobscot Bay in the mid-1980s. The characters: two men on a chartered yacht in northern Penobscot Bay. They have sailed up Eggemoggin Reach from Brooklin looking for an island — Great Spruce Head — where Fairfield Porter painted. They have been to his stunning posthumous show at the Whitney Museum in New York where they live and work, and somehow seeing the island will connect them with the part of their lives the show has touched.

WULP: I always ask myself, 'Why did I come here?' I've never understood it We were on this boat trying to find Great Spruce Head Island, but landed in Pulpit Harbor. As soon as I landed there I really had the oddest sensation. I thought, 'Oh my God, I'm going to live here,' which is really very odd to have that strong a feeling the minute you step into a place. So that fall I came back to find a house on North Haven. I had a very clear image of the house I wanted, but I found it instead across the Thorofare on Vinalhaven.

I then went through a whole process realizing I could no longer function in the New York theater. When I produced *Dracula*, it cost $300,000 and it had repaid its investment within one month's time. It then ran for 3 years after that with the two national companies and a company in London. That was no longer possible by the time I bought this house in 1985. I mean *Dracula* would have cost an enormous amount of money and would never have paid its investment back and you became increasingly dependent on the big theater owners in order to get financing. I just couldn't function in the New York theater the way it was anymore. The shows I was doing were just not making any money.

So I came here to Maine out of a sort of desperation, really. I owned this house, but I was broke. I had no money and I had no very clear idea how I was going to support myself, but I had nowhere else to go.

Working Stiff
THE SETTING: North Haven Schoolhouse near the center of the island. It is a redwood-sided building housing grades K-12, where some 60 island children scattered across 13 grades are clustered in multi-age groupings, once common in one room schools. As in most island communities, the school is the central institution of community life, where all its best hopes and worst fears come careening into view. Barney Hallowell, the principal, originally from a summer family, has worked in the North Haven schools for over 20 years. The origin of the island's drama program, he explains, resulted from observing many island kids' well-honed talent for mimicking their elders, often in wickedly funny ways. Hallowell recognized that his school was too small to excel in sports, but for a school to become great, it has to be good at something, anything; he sensed that a drama program might be the ticket.

WULP: When Barney Hallowell asked me if I would teach, in a way I grabbed at it, because I thought it was a way of improving my lot economically. I was working all the time. I worked at the lobster plant packing lobsters and I worked as a cook downtown when Phil Crossman had the Crow's Nest, because there was nowhere else to go.

There was a logic in Barney's offer, but I had never worked with kids at this level. And I didn't want to work in the theater again. I had had it. It was too painful.

I thought I was going to do something like Island Voices. You know, that we would examine the history of North Haven and try to create some sort of dramatic piece out of that material. But I really was a great failure at that. I got absolutely nowhere with that project with the kids. It seemed to me over and over again when I asked the older kids what they were interested in, it would be like ... there'd been a suicide in one of the farms over there ... anything with violence or suicide, they were just fascinated with that. I really didn't know what to do. I was more of a success with the younger kids....

Sacred Space
THE SETTING: A small boat crossing the Fox Island Thorofare. John Wulp, hunched up in a winter coat, is in the bow facing forward, into the bight of the northwest wind blowing hard over the water. Wulp is commuting to work at the North Haven School by crossing the narrow stretch of water separating Vinalhaven and North Haven. Although the distance is under a half mile, this narrow, windswept bit of water defines separate and remote worlds for much of the dark season of winter.

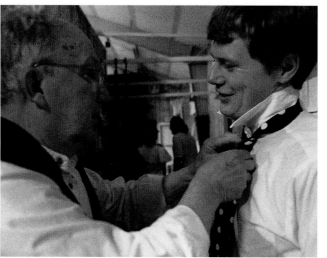
"I had never worked with kids at this level ... I really didn't know what to do."

WULP: I've always lived on an island. When we were kids we rented a house on an island in Lake Sunapee, and our whole family lived as a single unit. My grandmother, grandfather, uncle, aunts, cousins. I was lucky to have this aunt who organized masquerades and shipwreck parties, who was always saying, "Let's play at something." We lived in a great big house on an island and we were on our own for amusement. We would have musicals and pantomime. We would put up a sign to advertise. I just remember those times as being so incredibly wonderful. As a child it was tremendously exciting. I want kids on North Haven to have that sort of excitement that I had as a child.

Anyway, we did *The Enchanted Ferry Boat* with the third and fourth grades. The play was all about, if you could leave this island — what would you do? We would start by putting a groundcloth down and saying it is a sacred area. It's like going into a magic forest in which you are transformed. I tell this to my kids over and over again, that when they set up, when we put the ground cloth on the floor and they step foot into that area ... that [it] is like a sacred area and they must be aware that they are transformed at that moment. Cocteau says that a performer is like somebody on a high wire, and that if you laugh or cough or something, you can distract them and that they might fall to their deaths.

Anyway in *The Enchanted Ferry Boat*, certain of the

kids decided that they wanted eventually to return to that island and certain kids insisted that they never wanted to ... but it was very personal and very interesting. Certain of the kids wanted to see parents or relatives who had gone out of their lives. Later, some of the people on the school committee objected to that play, which just came as a complete shock and revelation to me. Some said we had made a play about divorce, which I don't think we had done at all.

Getting Earnest

THE SETTING: Calderwood Hall, a large, two story frame building just above the ferry landing on North Haven. Up the back stairway on the second floor is an improvised theater space with a small dressing room in the back, a simple set, and seating for a small audience crowded together on folding chairs. There is a audible buzz in the hall as backstage doors open and close, and you are aware of quickened footsteps amid hushed, excited whispers of the troupe, before the lights go down and the actors step into the set.

WULP: Then we did *The Importance of Being Earnest*. I was blessed with Marthena and Lydia Webster and Kim Lovell, Asa Pingree and then Seth Maxcey who came in later. But for some reason, they did not trust the material initially. I kept telling them that they ought to, and if they could just get it out the play would carry itself, and eventually, slowly ... they got it. I mean they said it! And I told them if they just did that, they would be all right; that the play was strong enough.

They articulated the play clearly, which made me question certain ideas about acting. I mean, you know about The Method and Stanislavsky and all that, but it seemed to me those were useless in a play like this; what was demanded was a clear articulation of the text. And they could do that.

Between Life and Death

THE SETTING: North Haven Community School at a school board meeting shortly after town meeting election, in which the balance of power on the School Board tilted to a majority opposed to the direction of North Haven's school program. In addition to the basic curriculum, this program provided for off-island school trips and an arts and enrichment program.

WULP: Then we did *A Midsummer Night's Dream:* just getting it on was a big success. This whole thing was threatened because I got very sick. I had two heart attacks, part of that was the stress of North Haven. The school board didn't appreciate what we were doing. I didn't think that [board], in any way at that time, valued what was being done. It was difficult to contend with; it was on your mind all the time; there was this intense pressure — all the time. So I had those two heart attacks. We were supposed to do *A Midsummer Night's Dream* in May, but it was postponed to August. So just physically getting it on was a victory.

What happened? Well, it took about an hour and 45 minutes whereas usually it takes about two and a half to three hours to put on. It moved with a speed and a clarity that was unusual. Again it was a matter of a clear articulation of the text. Just say it, let the play be heard. And it was a great success.

"I would like the kids to regard the theater as something holy and sacred."

There are people who have written letters saying the academic standards aren't high enough in the school and people shouldn't contribute money to arts and enrichment. I don't understand these people who make a distinction between academic standards and the work the kids are doing in these plays. There is no conflict. And the kids are doing this on their own as an extracurricular activity for the most part. So their devotion, coming night after night on their own time to these rehearsals and really working hard is extraordinary to me. Everyone always asks, 'What have you done to get these performances?' And I truly couldn't tell you because I believe the kids do it on their own

Now, nearly every kid in the school, except those whose parents won't allow them to, has participated. I would say 90 percent of the school has been in plays of one sort or another. And one year, we won the regional one-act play competition — and another, the state one-act competition. We've done very, very difficult material: Shakespeare, Anouilh, Thornton Wilder and Oscar Wilde. Most of the plays depend on language and the kids have learned to do this in a forthright sort of way. Certainly a study of the sort of plays we do would be demanded in any school with high academic standards.

I would like the kids to regard the theater as something holy and sacred, and that when they perform it's like this interchange between them and the audience that is, in strange ways, divine. I mean what is theater? We come together and for a moment we are joined as a community ... watching this thing in front of us that somehow defines our lives. Theater on the highest level is some sort of mystical experience. It's one of the few times when we become omniscient, like gods. You

"I came to Maine feeling myself a failure, not knowing where I was going to go or what I was going to do ... and now I am a man who, in his 70s, feels himself just coming into his full power."

know, we watch this experience and we can understand it. We can never do that really in actual life.

An Island in the Palm of My Hand
THE SETTING: Stone Farm in autumn's crimson light where John Wulp, the artist, has just finished a commissioned portrait of Sigourney Weaver and her daughter that will be shipped out in the coming week. Books, scripts, newspapers and canvases are scattered around and about in happy profusion. Remnants of a cook's deconstructed dinner party lurk in the kitchen. A Sunday *New York Times* is partially obliterated. Outside, a few late apples hang sweetening before they, too, harden off, and the limbs that now support them are stripped back to the bare essentials of winter.

WULP: I'm very proud of where I am in the painting. It seems to me that people's concentration has been shattered. You know there's this demand for speed. We want instant recognition. Speed, I think, destroys concentration. I'm sure it does. The Jackson Pollock paintings are wonderful, but they're like instant gratification, you know. The idea of sitting there with the brush and building up glaze after glaze after glaze, having the patience to do what real art requires; not many people have it. You can't have everything right away.

I'm no great proselytizer for art. That's the curious thing. I mean I hate amateur art. I hate the idea of art as therapy. Everybody seems to think they're going to be better off if they dabble in painting or write a novel or something and I hate to see that done to art because I think art is something that requires enormous dedication, [that] should require the development of a skill which is very often painful. Hard, hard discipline. It's not just something you do for the fun of it, I don't think. I think you do it with an enormous devotion.

You know, I do believe we are on some quest. We are trying to purify ourselves in some way, but we don't know how to do it. Camus said we are all trying to become saints without religion. Eventually there is very little that can be said. A person's life is what they do. It's a very lonely business. So making real contact with people is the only thing you can do. How do we become good? We don't find examples around us. It's very seldom that you find something that helps you. What is it that finally saves you — I think it is seeing things as they are — it's transcendentalist; it can't be gotten into words; it's too complex.

The High Wire
THE SETTING: Stone Farm. A thin line of woodsmoke etches the flat white light of frontal clouds in the morning sky. A woodstove inside throws enough heat to take the white-lipped bite off the edge of the morning. The living room of the farm is also its library, books lining the shelves.

WULP: Do you know Phillipe Petit? He is the guy who walked between the towers of the World Trade Center. There are things about him that just fascinated me. Like, he taught himself to walk on the tightrope by walking on the limbs of trees in his native France, and he'd go from one tree to another by walking on the limbs. Eventually he performed for Ringling Brothers Circus. And he hated it; he hated the commercialism of it. And he fell and seriously injured himself. So when I knew him what he did mostly is he would draw a circle on the ground. He would string a line between two trees in Sheridan Square and he would perform. He would juggle with fire or walk on the tightrope.

I thought Phillipe Petit was a great artist. I thought that sort of devotion to what he does and the skill with which he did it ... the skill was so enormous, to get out there on this high wire and not make a misstep. That's the thrilling moment. I think that great art, great theater art, is performed at that level of intensity. It's what Artaud meant when he said that a performer in a theater is like somebody signaling through the flames. I think that's what fascinated Hemingway about bullfighting, you know, that you have to put the blade in at the exact right angle to do it cleanly. We want performing at that level of skill. I think when you watch Chaplin at his very best you're aware of this extraordinary skill and you are breathless before it.

There are times when I think, 'Oh God, do I need to face another winter and those trips across the Thorofare?' But this has been an incredible experience for me because I have completely remade my life in this community. Yet I haven't in any way been deprived. I mean, we've had the association of somebody like Robert Indiana on *Red Eye of Love*. And Dana Reitz, whom I regard as a great dancer, is coming to work with the kids next week. For the Christmas plays, we're commissioning Gia Camolli to write the music and the daPonte String Quartet, which is a group here in Maine, is going to play that music. There is a continual stream of people that the arts and enrichment program brings into the school. So, it's not just me. Our program has solidified support for the school. I mean, I am amazed to find a community that values what I do. And none of this would have happened without Barney!

I came to Maine feeling myself a failure, not knowing where I was going to go or what I was going to do ... and now I am a man who, in his 70s, feels himself just coming into his full power. These things never would have happened if I hadn't come and hadn't found this spot. Ortega said the only genuine ideas in life are the ideas of the shipwrecked. "He who does not really feel himself lost, is lost without remission; that is to say he never finds himself, never comes up against his own reality."

Just Doing It
THE SETTING: North Haven Community Building, a short walk up the hill from the ferry landing. It's early April, but winter's chill clings to the water like a dark cloak. North Haven has struggled through a devastating community schism for two long years, pitting neighbor against neighbor and family against family. Recent elections, for the school board, and resignations, among the selectmen who opposed the direction of island education that the drama program came to represent, have changed the balance of power on the island to the status quo ante. But everyone knows nothing has really changed the deep dynamics of the underlying conflict, and maybe nothing ever will.

Into this deeply disturbing situation comes the school production of *Wind in the Willows*, a whimsical musical with a cast and production crew of over 50 kids drawn not just from North Haven, but also from Vinalhaven and Green's Island, to stage the show. A standing-room-only crowd of over 150 people is crammed into the rows of chairs and bleachers. Down front is a delegation from the New York theater world, including the playwright, Arnold Weinstein, whose play, although written a decade ago, is being performed for the first time here tonight.

WULP: *The Wind in the Willows* was quite a step for us because it was the first time we did an original play, that anyone entrusted us with an original play. It had never been performed before. We just did it and it worked. The kids were amazing. We used a whole new bunch of kids; we used mostly people from the lower grades. We still had to use some of the older kids like Asa and Chris Brown, but mostly they were kids just coming along.

When you see kids up there on the stage you cannot believe the kids are not getting a good education. But on the other hand, I have never been altogether able to do what I hoped I'd be able to do — give them enough confidence to feel that they could go anywhere and do anything they wanted to do. I think there is a fear, a certain naiveté, that is difficult to overcome. They are innocent of the world.

The End is the Beginning
The cast of *Wind in the Willows* is put together with a keen sense of the mythic identities that lie just beneath the surface of island kids, seemingly just waiting for a moment like this, for a director like this, to create the space and unlock their hearts. Ratty is as lyrical as the Mole is earnest. Badger is a high school youth who can seem to fill a door frame but who has not been on stage before Wulp came to North Haven. He is wonderfully solemn and ponderous in his portrayal of the ultimate enforcer among animals both good and bad. The slinking, black-costumed weasels and stoats are horrifically wonderful in their roles. The chorus of mice, with the heart-stopping solo performed by a flaxen-haired first grader, nearly brings the house to tears. But the show is stolen by the portrayal of the spectacular excesses of Toad, played by Jacob Greenlaw, a quiet eighth grader who works in the island grocery store. Jacob, like the play itself, is new to acting before a crowd of friends and island neighbors, new to costumery and lights, new to it all.

The production of *The Wind in the Willows* is an extraordinary display of the power of the arts to drive our worst demons back into the shadows, if only temporarily. Where the limitless energy of such a production comes from is unknowable, but the ability to find new meanings in such familiar territory is a kind of transformation. We search for this transformation in the sacred space of our lost childhood that we imagine, if only for a moment, we can recapture.

Philip Conkling is president of the Island Institute.

THE MR. WULP EFFECT

KAREN ROBERTS JACKSON

Nearly two years ago my two teenage sons, and thus, our family, became involved with an enigmatic man by the name of John Wulp. He, at that time, had a long list of adjectives, and a few expletives, trailing along with his name. He is an eccentric, demanding, omnipotent, retired New York City director who stages jaw-dropping performances by kids on North Haven island. We heard rumors about him on both sides of the spectrum: he broke kids' spirits, he asked too much, he was impatient, mostly he was intolerant of anything less than perfection. On the other side, lo and behold, he was extracting "perfection" out of junior high and high school students.

Our family also had a few oddities and adjectives of our own, such as the fact that we homeschooled our four kids on a small outer island — Green's Island, half a mile or so off Vinalhaven. I remember that we met Mr. Wulp (as all of the children call him) at a poetry reading at his home, Stone Farm, a sanctuary of art and literature. I also remember that in the kitchen was a table laden with the most mouth-watering, knee-buckling, ambrosia-of-the-gods desserts, created by this same quirky fellow. Sitting on his couch making small talk, I recall the woman next to me taking down a trophy from the mantelpiece and whispering to me, "Who the heck is Tony?" We were later to learn that, along with his "Tony," he had received an Obie and actually a long string of Broadway-related awards that impressed me, though I was fairly ignorant of their true significance at the time. Soon it came about that my two sons would work with him on his next production, *An Evening of Shakespeare: Sonnets and Soliloquies*.

My vague knowledge of the significance of a Tony or an Obie was about equal to my understanding of a Shakespearean sonnet or soliloquy. Mr. Wulp not only brought Mr. Shakespeare into our home, he brought him into our sleep, our dreams, our simple-minded understanding. Mimicking his eldest brother Dylan, my seven-year-old began spouting "...once more unto the breach dear friends!" My 11-year-old daughter memorized the lines of Lady Anne to better help her brother Tristan with his part of Richard III. For the weeks of rehearsal prior to the production (and every available occasion since), Shakespeare oozed from the pores of my children.

A lot of venting, seething and frustration oozed forth from my sons as well. This Mr. Wulp was not an easy man to please. He yelled at them, berated them, had them repeat lines and scenes to a point of exasperation. My husband and I had confided in Mr. Wulp that we were concerned about our 14-year-old's "dark side," that he was walking that pre-puberty line between good and evil. Mr. Wulp, already known for his uncanny casting ability, assigned Tristan the role of the nasty, hunchbacked Richard III. In rehearsal one day he brought forth his desired effect by telling Tristan, "I know you have a dark side, let me see it!" My son came home infuriated and humiliated, and I joined him by ranting around the house in my anger at this callous man. My attitude softened on opening night, when my two sons and their young male counterparts walked up to podiums in perfectly tailored tuxedos, four young women walked up in satin sleeveless gowns, and stunned the audience with their eloquence.

"This Mr. Wulp was not an easy man to please."

Thus began our unique relationship with John Wulp. I'm fairly certain that anyone who has had a relationship with Mr. Wulp would consider it unique. He extracts, demands, conjures a unique response, emotion, involvement with each person he meets. For better or for worse, he draws forth a full response from you, "casts" you, demands from you that which you might not have offered otherwise. I watched him draw forth a devotion and concern from my sons that I would not have thought them capable of in that phase of their youth. He could get them to cut, or not cut, their hair in the fashion HE prescribed. He would get my rebel sockless and shoeless son to wear shoes and WHITE socks by calling him up at 6:30 a.m. to remind him. He could get the boys to take out their earrings and the girls to vamp about convincingly with feather boas, circa 1920. He could demand that they give up not only their Saturday night carousing, but their Sunday afternoon as well. In some mysterious way he brought a discipline to their lives that a mother would die for. Mostly, he expected it of them, and they came through for him.

An example of the degree to which my sons would go for Mr. Wulp and a Mr. Wulp production is as follows. Living on Green's, my sons would head out in a rowboat to Vinalhaven. They would tie up at a friend's outhaul and walk a mile into the town of Vinalhaven to meet Mr. Wulp at the post office. From there they would drive the 12 miles to the other side of the island, call for a boat to come over from North Haven and take them across the Thorofare, where they would then walk up to the school for rehearsal. After two or three hours of rehearsal they would often stay several hours longer to work on scenery or the set or the sound system. They would then reverse the whole trip and end up at home a little before midnight. This was in all seasons, in all weather. To his credit, or his obvious insanity, at 69 and in poor health, Mr. Wulp's schedule was even more grueling than my sons'.

"In some mysterious way he brought a discipline to their lives that a mother would die for."

On these excursions my sons were entertained by the other side of Mr. Wulp; his singing, recitations, name-dropping of who's who on Broadway, literary debates, his probings and often childlike wonder at the world. They were influenced, equally it seems, by his impeccable genius and style — and his insensitive conduct and lack of grace. Through their association with him they learned to waltz and tango. They each got a turn feeling the weight of having lead roles. They shared the thrill of winning first place in the state high school one-act play competition. In addition, they were given glowing recommendations to colleges and student exchange programs by Mr. Wulp.

On one occasion I went to Mr. Wulp's home to interview him for a story I was doing. I was also hurriedly finishing a quilt I was piecing together for my son's 16th birthday. I asked if I could pin it together on his floor while we did the interview. Among his myriad talents, Mr. Wulp is also a (perfectionist) quilter. I was kneeling with my back to him, pinning the edges together, when Mr. Wulp said, "I don't understand this quilt..." I went into great detail explaining that the quilt was made from my son's boyhood T-shirts, and that each shirt was significant of an event in his life. When I turned around again, Mr. Wulp had gotten a seam ripper and was taking several of the panels apart. It seems that what he could not understand was how I could allow the panels to line up imperfectly with one another. He could not stop himself from rectifying the problem.

On another occasion, I mentioned that my son Tristan would be leaving for Ecuador just days before his 16th birthday. Mr. Wulp, surprising me, blurted out, "We must have party for him!" Though not feeling well at the time, just two days after the completion of one play and two days before beginning rehearsals of the next, Mr. Wulp hosted an extravaganza on my son's behalf. Some 40 people were invited to partake of turkey, ham, salads, potatoes and, of course, a beautiful birthday cake. In the midst of preparing the meal, Mr. Wulp's well ran dry. After everyone left, a neighbor brought over buckets of water for us to do at least the dishes. Soon Mr. Wulp and I were alone, me heating water on the stove, him doing the crossword puzzle out of the newspaper. Often in the past, Mr. Wulp had seemed disgusted with me for not knowing a certain author, or passage, or film, or set designer. This time, sounding weary from the day and lost in his own thoughts, he queried from the living room, "What has five letters and is something that goes a-leaping?" I answered back, "Lords, lords a-leaping." "Right," he said. In that moment I believe I felt the strange, childlike thrill that Mr. Wulp's young protégés must feel when they pleased the unpleasable one.

A Center in the Center

Extending back to minstrel shows and Legion Festival plays, North Haven has a long tradition of celebrating the arts. What the town does not have, however, is a center where students and community members can practice, demonstrate and exhibit their talents.

Now a community group is exploring an idea: renovate the more than century-old Waterman & Co. general store buildings on Main Street. A feasibility study has been completed. A theater design firm, Roger Morgan and Associates of New York, has drawn up preliminary plans to create a "community center in the center of the community," to enhance the programs of North Haven Arts and Enrichment. The island as a whole would benefit through revitalized downtown buildings, potential career opportunities and a place to showcase community spirit, pride, talents and accomplishments.

KIM ALEXANDER

(for more information contact Kim Alexander at P.O. Box 525, North Haven, ME 04853)

"As I haul myself up by the rope handles, the first thing I notice is how strenuous and yet meditative it is to hang from a cross."

"SOMETHING DIFFERENT TO LOOK AT"

MIKE GORMAN

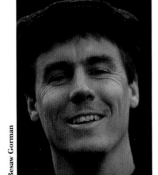

Mike practices his Euclidean Tai Chi on Main Street Vinalhaven.

IN 1898, ONE OF IRELAND'S GREATEST PLAYWRIGHTS, J. M. Synge, on the advice of W. B. Yeats, dropped his studies of French Literature and criticism in Paris. He journeyed to the Aran Islands off the coast of Galway, Ireland, "primitive and absolutely authentic places," to start describing a society with which he had some natural connection.

Synge's experience, captured in his book, The Aran Islands, proved to be an invaluable formative period for him and led directly to the writing of several of his greatest plays, Riders to the Sea and The Playboy of the Western World.

In 1988, on the somewhat less esteemed advice of an unemployed lawyer we had just met at a seafood market to "come out to Vinalhaven and make a mess of it," my brother Will and I undertook a journey of similar significance for us.

Despite the randomness of our source and its vague sense of delinquency we, like Synge, discovered an "absolutely authentic" place.

While Synge's experience on the Aran Islands can be read about in the beautiful book he wrote, neither Will nor I have ever written about our experience on Vinalhaven. Recently, however, I became inspired to write about it, after visiting the Aran Islands and re-reading Synge's book, which is based on the journals he kept on Aran.

Between the years of 1988 and 1996, as co-creators of the performance group, "The Fabulous Giggin' Bros.," Will and I wrote and performed plays, created an annual spectacle in the Fourth of July parade and staged various other dramatic events on Vinalhaven.

Beyond the performance of individual plays and dramatic productions, we engaged in an eight-year ongoing dramatic experiment, both hilarious and profound, with the community. Together, we challenged what we perceived to be the limiting idea of appropriateness — what you can perform, for whom and where — by pretty much performing anything, for anyone, anywhere we felt compelled to.

Our performances originated from a simple need to entertain ourselves, and grew to include those who felt a similar need to be entertained. What our audience knew, along with ourselves, was regardless of anything else, we were going to have fun.

From this perspective, we assumed nothing to be inherently valuable about our performances, beyond their value as entertainment. There was only the possibility of something of further value to be gained through individual interpretation. This approach was in keeping with the idiomatic expression of drama and beauty on Vinalhaven, as I experienced it, which was allusive and indirect. It would have been fatal to attempt too direct an approach in creating an authentic experience, just as it would be fatal to mistake the understatement of the island idiom for a lack of appreciation for beauty.

In general, the people of Vinalhaven, residents and visitors alike, seemed to welcome our adventurous spirit and attended our performances in consistent numbers no matter what or where we performed. We pursued our vision of theater and developed an artistic identity that would be recognized beyond Vinalhaven, at La MaMa Experimental Theater Company in New York.

While we gained our audience on Vinalhaven largely through absurd and comedic performances, it was our performance of Following the Northern Star, *a tragedy I wrote concerning a fishing community upon our return from New York in the summer of 1995, that was perhaps the greatest achievement of "The Fabulous Giggin' Bros." theater on Vinalhaven.*

From the reactions we received, it appeared to me that Following the Northern Star, *like Synge's* Riders to the Sea, *was embraced by the islanders in our audience as an authentic and moving representation of a tragic event they could imagine to be their own. The play revealed to me how intimately I felt we had come to know our island audience. The trust I felt they had invested in us and our Theater of the Absurd had finally paid dividends, in a dramatic performance that seemed absolutely real.*

To me, the most dramatic moments of my experience as a Fabulous Giggin' Brother on Vinalhaven did not happen directly on stage, but at its edge.

When, as a Jesus impersonator on the sidewalk in front of Boongie's video store after the Fourth of July Parade, I'm approached by two ladies who want to have their pictures taken with me.

When, as a fish broker on the back deck of The Fisherman's Co-op, minutes before performing in Following the Northern Star, *I'm approached by a lobsterman who tells me about the beautiful great blue crane he saw in the creek that morning, calling it simply "something different to look at, anyway."*

These moments of sharing most clearly speak of our most unique experience on Vinalhaven — the sharing of a language capable of expressing the potential for wonder, spectacle and revelation in every supermarket scene, coffee shop conversation and civic celebration, what I call the language of Giggin'.

Looking back, I would like to think that, if nothing else, the Fabulous Giggin' Bros., like the great blue crane in the creek, gave Vinalhaven something different to look at. I thank our audience for their adventurous spirit, their generosity and, most of all, their trust. We shared an experience I will never forget. You're in my stories.

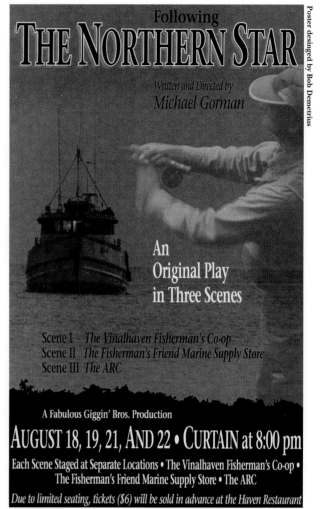

Poster designed by Bob Demetrius

August 16, 1995, 7:45 p.m., back deck of the Vinalhaven Fisherman's Co-op:
I'm in costume and getting in character for the 8 p.m. opening performance of *Following The Northern Star*, a tragedy which takes place in the mythical New England fishing port of Haversport.

I watch the last of the Vinalhaven lobster boats come in, and in my character's mind I wait for a Haversport trawler, THE NORTHERN STAR, to arrive. My character, the owner of a small commercial fish house, fears THE NORTHERN STAR may have gone down in a storm and the captain and his crew, good friends of his, may have drowned.

I pace the deck in my fish-scale-covered boots, my blood-and-gut smeared white coveralls and slime-shiny rubber gloves, stopping to alternately admire the ocean's arresting twilight beauty and curse its dangerous power. It's a walking wrestling match I'm engaged in, a physical struggle with conflicting feelings I have drawn from scenes I have observed on Vinalhaven.

June 9, 1989, Carver's Harbor:
It's 10 p.m. on a right thick o' fog June night. Seventy-two-year-old Curtis Connover emerges from his aluminum trailer, perched 200 feet above the harbor. Curt's a sweet, tender soul who loves the ocean and would die if he were removed from the sight of it. But tonight he's alight with anger. He's remembering the night, just a few weeks ago, the sea took his only son from a fishing boat.

Tonight he's got his boxing gloves on. If he doesn't kick the ocean's ass, he's at least going to get a few things straight. Curt's been drinking, and shoulders his way sluggishly through the battered screen door in his navy blue captain's hat and pea coat, staggers down his three unpainted wooden front stairs and weaves an erratic but determined line to the harbor's edge. He stands on the end of a dock and takes his first swing.

"Come on, you son of a bitch!"

His arm flies out wildly from his side and nearly carries him off into the billowing tide.

"What more do you want from me? I gave you 40 years on the boat! Six more in the navy! You took my brother! And now you take my only son! What more could you possibly want from me?"

Curt hears his words absorbed by the fog and the water but not returned. He takes another swing at the ocean's cowardly lack of response, "Come on, you son of a bitch, I'll take you on!"

August 12, 1992, Harbor Wharf:

It's one o'clock on a sultry Saturday afternoon. A polite and respectful crowd has gathered at Blanche Bowdoin's art gallery to view her latest paintings, which exclusively depict the life of the harbor; in particular, the fishing boats. Most of the viewers drink wine, eat cheese and simply admire the paintings. A few discuss the possibility of purchasing or bartering for one. The atmosphere is very relaxed, with no real sense of urgency.

Ambrose Wheeler leaps from the rail of the herring seiner he has just come in on, races across the dock and lurches up the stairs to Blanche's gallery on wobbly sea legs. He knows Blanche has made a painting of his old boat, THE CAPTAIN'S PRIDE, and is seized with the fear that someone has beaten him to buying it. He crashes through the door, reeking of herring, cigarettes and coffee brandy. His eyes scan the walls desperately for the painting as he stalks into the center of the gallery, his hands pulling cash from his pockets.

Will Gorman as the baby Jesus with Bunny Beckman as the proud mother Mary in the Vinalhaven version of 'The Greatest Story Ever Told."

"I want that painting!" he yells, and inadvertently casts a crumpled hundred dollar bill through the air as he spots the painting and points in its direction. Three more hundred dollar bills fall from his pocket onto the floor. The crowd parts as Ambrose steams for the painting and pulls it from the wall. His knuckles clench on the frame with the grip of a drowning man as he stares into the painting with both anger and trepidation - the victim of an assault facing his assailant in court.

"That boat tried to kill me!"

He looks around, wild-eyed, to see if this fact is registering at all with anyone in the gallery, before being sucked back into the painting. "I'm not kidding, she tried to kill me! Middle of a storm. Blowing a gale. Scaled me over to one rail and then scaled me back again and over the other rail, tits to the stars!" He looks around again to see if anyone else is seeing the ocean's horizon upside down.

Back deck of the Co-op:

As I watch the last lobster boats pull up to the dock and unload, and I get ready to enact the drama of the boat that hasn't come in yet in the play, I enter a surreal time.

August 15, 1995, enroute to The Fisherman's Friend:

This is our last rehearsal — our only rehearsal in the physical space of the Co-op and The Fisherman's Friend, the locations for the first and second scenes of Following the Northern Star. I hop into the back of the pickup truck and we tear out of the Co-op parking lot and head a hundred yards down the road to The Fisherman's Friend. We need to practice the necessary haste to beat the audience to the scene tomorrow night. I feel intensely exhilarated, the way I always do, traveling 40 miles an hour in the back of a pickup truck, within spitting distance of the ocean on a moonlit night. More than that, I feel my perception piqued by how strange everything has become.

Still seeing things from the perspective of my character, following our rehearsal of the first scene, I don't recognize the street we are traveling on as the street I travel on everyday, but the street to The Outrigger bar in the play. The cars we pass I don't

Vinalhaven residents perform in "Biffing Mussels" at LaMaMa E.T.C. in New York City: (left to right) Bill Chilles, Will Gorman, Mike Gorman, Marguerite White, Peter Farrelly.

recognize as the same cars I've passed for the last three months. Even the grass, the trees and the stray dogs make me feel like I'm someplace else. It's like watching a movie that has come to life.

Back deck of the Co-op:
This is the experience I have envisioned for the audience, but have heard some doubt that the audience will suffer the inconvenience of traveling to three different locations. Will people really not come because they will have to get up out of their seats three times, I wonder. Will they really not come because they will have to sit on creaky wooden bleachers, have to inhale the fish-funk smell of the Co-op or suffer the light of the moon to guide their steps between locations? Will people really view the journey as an inconvenience rather than an adventure? I remember other times when seeds of doubt, planted in my head, have made me question my intuition.

July 3, 1994, preparing a performance for the Fourth of July Parade:
"You're really going to be Jesus on the cross?!" a concerned friend asks.

"Yeah, why not?" I respond. "The theme is 'books on parade' and we're doing 'The Greatest Story Ever Told.' How can you portray the story of Christ without the crucifixion?"

"I can't believe you're going to do this," my friend persists.

"Why not?" I really haven't imagined anything other than an enthusiastic response from the Fourth of July crowd.

"It's ... well, it's blasphemous," stammers my friend, looking a bit blasphemed herself. "Jesus on Main Street? It's going too far."

July 4, 1994, Main Street, on the cross:
The cross is a beauty, made out of termite damaged six-by-six beams from a dismantled 18th century barn. Not strong enough to reuse for the purpose of spanning a floor in a new barn, but plenty strong enough to support the weight of a 172-pound Jesus impersonator and half a dozen nails. Actually, we hedge on the realism and use rope instead of nails, and as I haul myself up by the rope handles, the first thing I notice is how strenuous and yet meditative it is to hang from a cross.

The effort of keeping my head raised and arms outstretched is like doing yoga in one angular position. "Euclidean tai chi" might best describe it. The second thing I notice is how quiet the audience has become. I want to think the pin drop quiet is in reverence for the speechless beauty of the act, but the voice of my skeptical friend rings in my head, "Blasphemy. You've gone too far."

"No, it's beautiful," a voice whispers back.

A dialogue starts in my thoughts as I watch my lawyer and his father, masquerading as the Devil and God, joust below me in the eternal battle of good and evil in our colorful pageant. The voices in my head raise from a whisper to a shout as "Sympathy for The Devil" plays on the devil's boombox and "Jesus Christ Superstar" plays on God's.

"Blasphemer!"
"Beautiful!"

The crowd remains silent. A nervous flush rises in my body and I step down from the cross. I topple the 10-foot, 150-pound crucifix onto my shoulder and drag it down Main Street as fast as I can. Near the end of Main Street two elderly women whom I recognize as members of the Unitarian Church approach me excitedly.

"Oh no," I think, "This is it, they've come to cast the first stones." I cringe as one of the women reaches in her pocketbook and pulls out ... a camera!

"Could we get our picture taken with you?" asks the woman with the camera.

"It was just too beautiful for words," says her companion. The cross suddenly feels much lighter on my shoulder.

August 13, 1994, **The Wind,** *the island weekly newspaper:*
We have published our ad for this week's play completely in Spanish. There was no practical reason to print the ad in Spanish. Our play has absolutely nothing to do with Spanish and, as far as I know, the Spanish-speaking population of the island is very, very small. We did it on a whim, just to give people something different to talk about. We've come to view our ads in *The Wind* as little verbal joy rides for shut-ins, providing entertainment for those who can't attend the event as well as making the practical announcement of the time and place for those who can.

Mike schools Will on the finer points of racing pigs at LaMaMa E.T.C.

```
UN NOCHE FABULOSO
CON LOS HERMANOS FUBULOSOS
      DE GIGUENDOS
Los Hermanos Fabulosos de
Giguendos presentan: "Death
by Joinery"!! En un noche
de teatro fantistico con
una otra producion
superlativissimo "The Final
Scene" por Maria Eder y los
actores que no conozen que
se llaman. El ♦ ⚓ 6ᴛᴴ De
Septiembre, 8 pm en el
ARC.  Vamos a giguer como
campeones!  Que Fabuloso!
*************************
```

August 13, 1994, the Tide's Edge Restaurant:
It's 5 a.m. Peter Henderson picks up his *Wind* after Jim Stevens picks up his, and Tudor Calderwood picks up his from the box outside the entrance to the restaurant. The three lobstermen sit down at a table and wait for their coffee. Tudor Calderwood opens his *Wind*.

"Chrissake, would you look at this!"

"Chrissake," says Peter, looking at the same page in his *Wind*, "It's all in Spanish!"

"Chrissake!" says Jim.

August 13, 1994, the Tide's Edge Restaurant:
By 5:20 a.m. Tudor Calderwood has turned the restaurant into a high school language lab.

"Escuche y repetan," he instructs, and points to his cup of coffee on his way out the door to his lobster boat. "El cafe es muy caliente!" "EL CAFE ES MUY CALIENTE!" The rest of the early morning diners sing out in chorus.

"El agua as muy fria!" Tudor tutors, pointing to the harbor.

"EL AGUA ES MUY FRIA!" repeats the class.

August 13, 1994, Calderwood Neck:
Jane and Aubrey Beal get up from their kitchen table and make their way carefully across the granite ledge to the foot of our ladders. They have been watching us paint the house next door to theirs for most of the morning and have come over to get acquainted. Jane and Aubrey are very old. Jane has trouble walking and Aubrey has trouble seeing. They cling to each other and make better progress together than either could make alone. Their dependence on each other and their persevering spirit bring to mind the characters of Vladamir and Estragon in Samuel Beckett's play *Waiting for Godot*.

"Good mornin'," says Jane, peering up the height of our ladders.

"It's a beauty," I say down at their upturned faces.

"Stunnin'," says Aubrey.

"Say boys," says Jane, her voice and smile as warm as the August sun. "We just wanted to let you know how much we appreciate those ads of yours in *The Wind*."

"Right corkers," says Aubrey. " 'Specially that one in Spanish."

August 13, 1994, town:
Armando, a New Yorker of Spanish descent, is furious. He paces the kitchen of his mother-in-law's summer home, his Latin blood on the rise. He slaps the four-page photocopied island newspaper with the back of his hand as if he were slapping the face of his adulterous wife in an old-time Spanish melodrama.

"What is the meaning of this?! 'Los Hermanos Fabulosos! Que Giguendos!'" he quotes. "Giguendos is not even a word! Why do they do this? It is not their language."

August 13, 1994, Calderwood Neck:
Word of Armando's dismay at what he perceives to be our vulgar appropriation of his native language reaches me on the ladder via a mutual friend.

"Have we really offended him?" I wonder, the old doubt creeping in. "Have we offended anyone else? Will people not come to the play because they can't understand the ad?"

August 15 and 16, 1994, The A.R.C. (Arts and Recreation Center):
On Friday night, 140 people pack the A.R.C. for the performance of our play. On Saturday night, 130 more attend. My doubt is dispelled. Apparently, our Giggin' has lost nothing in the translation.

The back deck of the Co-op:
I can hear the people entering the Co-op to see tonight's play, scuffling across the raw board floor and climbing into the bleachers. Their voices echo off the naked stud walls and collect in one body and volume as more and more people pour through the raised garage door. I recognize many of the voices and I imagine their faces. I can picture where people are sitting and how many have come.

There's something miraculous about hearing an audience gather. They just start appearing and I can't quite believe that all these people have come. Here in the Co-op, as at La MaMa experimental theater on the night of our first performance in New York four months earlier, I experience this miracle in its most immediate sense. Like the now famous "One Night Stands" at La MaMa where founder Ellen Stewart welcomes young artists to perform on her stage for a night and invite their friends, this is where it all begins. And this is what it must always come back to, and what can never be destroyed if you want to have an authentic experience.

The back deck of the Co-op:
It's 10 minutes to show time. I wait anxiously to begin the performance of *Following the Northern Star* and am surprised to see a boat coming in so late.

August 16, 1995, The harbor:
At the same time, Luther Tolbert looks up in surprise at all the bright lights on in the Co-Op before he cuts the wheel to his boat and swings her up alongside the dock. "What the—," he says out loud, trying to figure out what all the lights are about.

"Looks like a freakin' Christmas tree."

"Giggin's got a play on," says Jamie Thompson, Luther's sternman, who jumps over the rail to tie up the boat, eager to unload and get to the show.

"Friggin' Giggins," he says, sounding annoyed, but unable to suppress a curious smile at how different the Co-op looks.

The back deck of the Co-Op:
I watch Luther lurch his way up the walkway, and begin to prepare some sort of timid, apologetic greeting to explain my presence here. But then I think, the hell with that. Tonight I run this place, and if I'm to convince any of the 80-some-odd people waiting inside on the bleachers of that, I better start convincing myself right now.

"'Sgoin' on, Cap?!" I greet Luther robustly as he steps off the walkway and onto the deck.

The back deck of the Co-op:
Luther stops in his tracks, unable to believe what he's seeing; an actor playing a fisherman in the Co-op, and a bleacher full of people getting ready to watch him. He feels completely displaced. He's surprised not to feel resentful or suspicious but strangely relieved, as if the actor's presence has lifted the burden of work from his shoulders. He doesn't know what to say, but suddenly feels free to talk about something different than fishing. He thinks of the sight of the Co-op from the harbor, all lit up, and then he remembers something he saw on his way to his boat this morning.

"You know," he says, working around to it after talking about the moon and the air and tomorrow's forecast for a little while, "I was on my way to work this morning ... the sun was just coming up on Indian Creek ... and I saw this great blue heron ..."

The house lights snap off and I hear the music start. I wait for Luther to finish. This is a dramatic moment. He's telling me something important — handing me a little note to give to my character on stage that night. "It was ... Well, it was kind of like comin' in and seeing you fellas here with the Co-op lit right up. Well, you know, something different to look at, anyway." Just then, the stage manager pokes his head around the corner of the door and says, "You're on."

Mike Gorman and his brother Will are preparing for a fall production date at La MaMa E.T.C. in New York.

THE DREAM KEEPER

Ashley Bryan of Little Cranberry Island

SUSAN HAND SHETTERLY

Photographs by Bridget Besaw Gorman
Drawings by Ashley Bryan

WITH HIS HANDS deep in the pockets of his corduroy jacket, Ashley Bryan walks along a dirt path on Little Cranberry Island that leads to Hadlock Cove and the boat dock. It is almost four in the afternoon, time for the mailboat from Northeast Harbor. November winds blade through the tall spruces on either side of the path, and beyond them, the water of Great Harbor is beginning to whitecap. The big hills of Mount Desert rise darkly against the sky. There is no sign of the boat.

Bryan is reciting a poem as he walks, his voice mining the music in it, as if it were almost, but not quite, song:

"The night is beautiful," Bryan intones.
"So the faces of my people.

"The stars are beautiful,
So the eyes of my people.

"Beautiful, also, is the sun.
Beautiful, also, are the souls of my people."

The poem was written by Langston Hughes in 1922. Like most of this poet's work, it is rooted in African-American experience. One might think it an unusual choice. here, on this island. One might expect, perhaps, something closer to Robert Lowell's "The Quaker Graveyard in Nantucket" or a poem by Philip Booth that celebrates the elegant reserve of coastal language. But Bryan, reciting Hughes, extends the meaning of the words. The poem, for him, speaks to all the people with whom he shares community. And these include the people of Little Cranberry.

Great Cranberry and Little Cranberry, the largest in a five-island archipelago, were settled by whites in the 1760s. Names in the early history of both islands — Spurling, Gilley, Stanley, Bunker, Young — are familiar in Hancock County today, and a number of island people trace their ancestry back to those first settlers. The settlers chose the islands because the combination of farmland, pasture and fishing grounds offered a generous living for hard-working people. Here they built homes and schools and boats, collected taxes, drew property lines and gave stipends to the poor.

"Community," Bryan says, "means a sense of who you are. You are — yes — dependent on others here. But a part of being in community is the independence it gives you. If there's someone who doesn't want to be part of it, people will understand that. You can live as a hermit if you like.

Ashley Bryan paints with tempera and gouache, carefully opening up his designs so that they shine with light and yet retain an almost Byzantine quality of abstraction.

"But when I came to the island, I touched on the sense of community immediately. If you get off the boat with a package, you don't have to struggle with it. It will pass to one person. It will pass to the other person. It will be a chain of hands. And it has nothing to do with what they think of you — it is reaching out in terms of this sense of community.

"Growing up in the Bronx, I was aware that the city itself breaks down into small communities. Small communities in big New York. And that's what I found familiar here."

•

He visited Acadia Park in 1946, looked across the water to the Cranberry Islands and said to himself, "That's home."

He was a 23-year-old graduate of Cooper Union and Columbia University, a veteran of the Second World War and had just spent the summer painting landscapes at the Skowhegan School of Art. For the next 30-odd years he rented summer places and studios until he bought his own house in 1974 in the village of Islesford on Little Cranberry. It gives him, he says, a quiet center.

"There are certain unchangeable things about a geographic place like this. They can't built skyscrapers. You have a modest placing of homes. I'll put down my work and go out for a walk and come back fresh. When I'm here, time is so big — it feels so big to me."

Bryan is a writer and illustrator of children's books, a Professor Emeritus of Art at Dartmouth College and a landscape painter. Over the years he has won various commendations for his books for children, including the Coretta Scott King Award. He explains that until the end of the 1960s, most children were reading books and seeing illustrations that were exclusively white: white faces, white narratives. With the creation of the King award, things began to change.

"It opened up the field," Bryan says. "There are stories now about Native Americans, and people with roots in China, Japan, Korea, India — many countries. Of course, that's what the United States is. Children should know about others among whom they are living, or who live in the country with them, before they get to other parts of the world. They should know American society in terms of these stories and these lives."

Some of his books, such as *Sing to the Sun, What a Morning, What a Wonderful World* and *The Story of Lightening and Thunder,* are brilliant celebrations of color that find their centers in African and African-American experience. He paints with tempera and gouache, carefully opening up his designs so that they shine with light and yet retain an almost Byzantine quality of abstraction. They look somewhat like stained glass windows in full sun. This is not surprising, given the fact that, as a small boy, Bryan attended a German Lutheran church in his Bronx neighborhood, and what he loved most were the Gothic windows.

It was, he says, a few years before he noticed that he and his family were the only Black people at the services. Eventually, he became a Sunday school teacher and an active member of the church's outreach programs. When the building caught on fire 10 years ago and one of the beautiful stained glass windows was destroyed, Bryan designed another to take its place. By that time, the neighborhood had changed, and the congregation was mostly Black and Hispanic; the window he created for them features a Black Christ. "I made it all color, with no etching for volume," he explains. "A Black Christ, and color against color against color all the way through."

•

"A book," says Bryan, "will stir a child to do something on his or her own, such as learn more about the lives of great Black artists. When I wrote and illustrated *What a Wonderful World*, I wanted to introduce Louis Armstrong to a generation of children who might not know him.

"I start all my programs reading from the Black American poets, then I read from a story of my own, and the connection is clear. The prose of my stories is based on this poetry. The children and the librarians may not have thought of these poets, might never have heard of them. As I read them, they see how accessible

Bryan has kept alive the dreamlife of his own childhood.

they are, how much a part of everything of childhood they are. Afterwards, they come up and write down the names of the different poets. They bring their books into the libraries. They include them. That means a lot to me.

"I am trying to make the sound of the voice in the printed word very clear so that when you are reading, you are listening, you are hearing the story. Poetry is an art of performance and you must spend time preparing. I prepare readings from poets and that allows me to explore the dynamic range of the voice, and it's that range that's present in my stories. I connect the pleasure of the voice with reading a book so that children will become readers for life.

"My sources are poetry and the oral traditions. The origins of all stories are in oral traditions, maintained and memorized because of the play of language and sound — the patterns. They are in every culture and every family. And I think the deprived family is the one that doesn't tell its stories.

"Children love to hear stories, but they also like the company, someone who cares enough to tell them a story or to read to them. There's some kind of comfort that we need as human beings that we get from this."

•

Ashley Bryan's house looks much like any other on the island, a subdued, two-story shingled structure, but step inside and one enters a fantasy of childhood — not my childhood, and probably not your childhood, nor anyone else's, unless they were raised by an alchemist.

Rather, he has created a space that celebrates the imaginative reach of children. It is not a play house, but adults and children, stepping inside, can't help but respond to the capacity of the human mind to invent and disguise.

The rooms are filled with shelves and tables and cabinets over which, neatly arranged, sparkling and dustless, toys from all over the world are lovingly displayed. They crowd together as if they came out of a fitful dream: all sorts of animals made of wood or cloth or metal, puppets of every description, dolls, cars and motorcycles and tiny houses. On shelves against the wall, hundreds of children's books are stacked. In the bookshelf under the stairway to Bryan's studio sits Rampersad's definitive two-volume biography of

Langston Hughes. Next to Bryan's bedroom, on a sheet music stand, rests a picture of Louis Armstrong, and on a chair by his kitchen table lies a book featuring on its cover the elegant, brooding face of Paul Robeson. There are all sorts of ghosts here, and relics from all sorts of dreams — those attained and those deferred.

Four large boxes contain Bryan's collections of bones and beach glass, the glass separated by color into smaller boxes: handfuls of smoky blue and emerald green, browns that once were parts of beer bottles and now look more like the subdued, direct gaze of someone's eyes and the clear glass, etched by the tide so that it looks permanently scrawled with snow.

The bones he has collected are from fish and birds and mammals. They are white and clean, and beautiful because they are shaped for function. Out of these pieces of flotsam he finds along the island shore he creates his stained glass windows and his puppets.

The stained glass panels he has made from beach glass and papier-mache tell the story of the life of Christ. The puppets, however, come out of older mythic sources. There is Owl Man, fuzzy-jowled and yellow-beaked, tall, thin, wingless, dressed in rags. There is Boneface, mouthless, earless, staring rigidly ahead with blazing red eyes of glass. And there is the elephant king, rattling the bones of its dead.

Bryan is a man who has responded not only to the dreams of people such as Hughes and Robeson, but has kept alive the dreamlife from his own childhood. He seems, at 75, a man almost out of time — or beyond time — and uniquely accessible to today's children.

He travels all over the country lecturing and reading from his books. One of the most valuable contributions he has made to the state of Maine is the time he takes to visit classrooms and libraries to read stories and to talk to children. For a few of these children he is their first live contact with an African-American. In this simple and profound "getting to know you," Bryan deepens their lives: he comes to them as a storyteller, and as an American beyond their immediate experience. And they don't forget him.

"I want people to know that what is offered by one individual is open to all people," Bryan says. "If you look in terms of color only, you miss the point. If you love the work, you make it your own, and offer it to others."

•

Bryan's parents immigrated from Antigua before he was born. He got to know their island as an adult, and he now has a family home there which he visits occasionally. But something equatorial seems to have always pervaded much of his work.

In one of his poems, he wrote: "I dig/To Africa." This direction of digging is apparent in the drenching sun-bright colors he uses, and in the linoleum block prints he has made to go along with the retelling of African tales. Although the prints are often only in black, they have the strong, energetic boldness reminiscent of indigenous African art.

Yet Bryan asks us to see the colors of New England, to find the boldness and the heat in them: "It is," he says, "the most local things that are universal. When I'm painting in the garden here, people ask me if I've done this work in the Caribbean. People don't know how colorful New England can be. How rich it can be. They think grey and barren. But I see color."

•

Neighbors on the island are always dropping in to visit and to chat. Bryan shares a history with them that goes back years. Ted Spurling, who has made his living as a lobsterman and merchant mariner, is the island historian and an old friend of Bryan's who now contributes a monthly column to *The Working Waterfront*. Spurling writes that Little Cranberry " ... is a place where a man can still walk in friendship with his neighbor."

"I am a part of where I am," Bryan says. "I work into whatever is going on, and here I am one of the island people. My kind of toughness is what I do in my work, in drawing and painting. People respect that, as I admire and respect what they do."

He likes to tell about the time he gave an American Library Association lecture in New Orleans, and took a friend with him, a retired island school teacher in her late 80s. He says she stole the show. "They loved her! She's a born storyteller," he says with obvious delight. "She has that poise, that way of speech. Everything she says is story."

"In the summer I paint in the garden," he says, standing in his studio looking out the big window filled with light. It offers a view of his flower garden, now gone to seed, and beyond it, a horizon of sky and big trees and a hint of ocean. "I'd be out in the garden and one of my friends, a fisherman, would pass by day after day, year after year — for about seven or eight years. So one day, he comes over, stands by me and says, 'Hmm ... still painting the same picture?'"

Bryan throws back his head and laughs. "That's just so wonderful!" he says.

His studio is stacked with canvases, paintings of the flowers he has found on Little Cranberry. This style is loose and free, the outlines indistinct, as if a constant wind were blowing through the flowers as he painted them.

"My neighbor, Emerson, died last year," Bryan says. "He was 88 years old. I'd bring him his mail and come over in the evening to watch the news with him. He was a farmer. He had rows of dahlias. Whenever the ladies

Little Cranberry, writes Bryan's neighbor Ted Spurling Sr., "is a place where a man can still walk in friendship with his neighbor."

would stop by, he would cut a bouquet of flowers. He was a wonderful man. The dahlia man. I paint those dahlias of his now.

"And I spend a lot of time walking along the shore. That's where the life is: the open patterned woods, and the things that wash ashore in winter. Sea smoke rising. It's always new. It's never anything you have seen before. It has to be discovered."

The mailboat swings out in front of Sutton Island, cutting a wake through the Gilley Thoroughfare. Bryan pulls up the collar of his jacket and watches the boat. The weather is turning, and the lobster boats in the harbor are starting to shift at their moorings. Bryan stares out at the water and he appears, suddenly, iconographic: a northern man facing into the clean cold start of another winter.

In *Sing to the Sun*, his book of paintings and poems published in 1992, he has written a poem entitled "The Artist." In it we catch a glimpse of the harder edge: that private toughness that can turn the stuff of life into something worth saving — the gleam in a puppet's eye that comes from a piece of broken glass.

Susan Hand Shetterly *is author of* The New Year's Owl *and other books.*

I know a man
Like a child
He loves to paint

> **He can paint anything**
> **He sets his heart to**

He knows
That to have
Anything he loves
He can have it
Fair and forever
If he paints
A picture of it

> **He knows**
> **That to face**
> **Anything that hurts**
> **He can do it**
> **Transform the sorrow**
> **If he paints**
> **A picture of it**

This is how he lives
This is what he does.

A Siren Call

For John Fowles,
islands are the most powerful metaphor of all

KATHERINE TARBOX

ONE OF MY FONDEST MEMORIES is of a wild day last May when John Fowles and I - despite my protestations - took the LAURA B out to Monhegan in 12-foot seas and a bitter southeast gale. There he was, the great English writer, now encumbered by age and unsteady legs, holding on for dear life as the boat rolled side to side, the oranges he'd bought for the trip bobbing around his ankles in the water we'd taken on. He simply had to see the bird migration and stalk elusive orchids. His profound love of nature and of islands always vetoes mere reasonableness.

John Fowles has been enchanting readers, whose many lives he has touched and often changed, for nearly half a century now. His novels are deeply seductive and mysterious stories designed, as he says, to draw his readers into a confrontation with their naked, elemental selves. The metaphor on which all these magical novels turn is "island-ness" - a complex thought-feeling drawn from his association with actual islands.

On the Greek island of Spetsai, where he taught as a young man, he learned that the seeming solidity of one's identity is undermined by the experience of being islanded. As he says in *The Magus*, the novel based on the Spetsai experience, "It was like being at the beginning of an interrogation under arc-lights. Already my old self began to know that it wouldn't be able to hold out." Next, he wrote *Islands* (from which the following passages are chosen), about the Scilly Isles, whose ancient mysteries beset his imagination. Most recently he has been attracted to the wilderness of Monhegan where, despite the fact that his "righting" ability had been impaired by a stroke, he has wandered over the difficult bluffs and cliffs looking for wildflowers and warblers, and sat atop Whitehead transfixed by the croodling eiders. He's called to his totem bird, the raven, who stubbornly refused to show himself. He has studied the many paintings of the island and admired the courage of the artists who have tried to capture its essence. One fine afternoon he made a rough crossing to Southern Island to visit with Jamie Wyeth, to speak with him about art and the magic of islands.

Fowles has found on the Maine islands the same "living feeling" he's found on unspoiled islands all over the world. These islands, for Fowles, can only be understood in counterpoint to the mainland, the vast continents that have been marked and defiled by the depredations of capitalist greed, pointless aggression and gross stupidity. Mainland is, for him, a metaphor for a sort of hell that supports all the excesses and insanities of a technologized world. Island is the symbol for all we've either lost or forgotten. Island is virtually synonymous with Nature itself, the existingness of all things, the "naked reality," home of silent language and the pagan mind.

Fowles is grieved not just by what we have done to Nature, but by the ways we have travestied our own human nature, by allowing ourselves to become subject to relentless cultural brainwashing. We've forgotten to ask the essential human questions Fowles hears in every dove's song: "Who are you? Why are you here?" Islands answer to what is natural in each of us. On islands the noise, busyness and flickering signals of mainland reality fall away, while stillness and centeredness return to us, wherein we might once again hear the clarity of the inner voice. Everything Fowles writes testifies to this necromantic ability of islands to take us apart, set us adrift from cultural moorings and to strip us to essentials. To him, all islands issue a siren call that both bewitches and destroys, but the island causes a necessary dismantling. It unsettles us, melts down our false identities and pretensions, purges and cleanses all that is overcivilized in us.

John Fowles's life on earth has drawn him along the arc of this one destiny: to explore courageously the island-ness of each human being and its simultaneous and perplexing oneness with the mystery that is Nature. He has been exquisitely moved by the perpetual vivid miracle of being itself. Always he is shocked and humbled by the power of Nature, and, he says sadly, "It pains me that I can say no more about all that has said so much to me."

Author of The Art of John Fowles, **Katherine Tarbox** *teaches at the University of New Hampshire.*

John Fowles' novels include *The Collector, The Magus, The French Lieutenant's Woman, Daniel Martin, Mantissa* and *A Maggot. Wormholes*, a book of essays, was published in 1998. The selections on these pages are drawn from *Islands*, a book of reflections published in 1978.

I first got to know of Monhegan through the kind agency of a formidable Maine mother, a gifted lecturer and teacher in New Hampshire, Katherine Tarbox, and almost at once recognized various key symptoms of what I will call "islandness." I had been here before, I knew these feelings, this almost metaphysical sense of isolation. I was first lured to Monhegan by the promise of a very rich bird migration, which the island in no way betrayed. I've been a birdwatcher all my life and found myself in paradise. I loved the close-community feel of the island and then its croodling male eiders close offshore, to say nothing of so many much rarer creatures, both avian and human. There was a poetess, a most pleasant surprise, and the kwarking sound of ravens, for me the indispensable sign of greeting from all truly wild and remote places. It is a sound I have sought all over the world. On Monhegan I knew I was home.

Photographs by Peter Ralston

True islands always play the sirens' (and book-makers') trick: they lure by challenging, by daring. Somewhere on them one will become Crusoe again, one will discover something: the iron-bound chest, the jackpot, the outside chance. The Greek island I lived on in the early 1950s, Spetsai, was just such a place. Like Crusoe, I never knew who I really was, what I lacked (what the psycho-analytical theorists of artistic making call the "creative gap"), until I had wandered in its solitudes and emptinesses. Eventually it let me feel it was mine: which is the other great siren charm of islands — that they will not belong to any legal owner, but offer to become a part of all who tread and love them. One's property by deed they may never be; but man long ago discovered, had to discover, that that is not the only way to possess territory.

It is this aspect of islands that particularly interests me: how deeply they can haunt and form the personal as well as the public imagination. This power comes primarily, I believe, from a vague yet immediate sense of identity. In terms of consciousness, and self-consciousness, every individual human is an island, in spite of Donne's famous preaching to the contrary. It is the boundedness of the smaller island, encompassable in a glance, walkable in one day, that relates it to the human body closer than any other geographical conformation of land. It is also the contrast between what can be seen at once and what remains, beyond the shore that faces us, hidden. Even to ourselves we are the same, half superficial and obvious, and half concealed, labyrinthine, fascinating to explore. Then there is the enisling sea, our evolutionary amniotic fluid, the element in which we too were once enwombed, from which our own antediluvian line rose into the light and air. There is the marked individuality of islands, which we should like to think corresponds with our own; their obstinate separatedness of character, even when they lie in archipelagos.

Island communities are the original alternative societies. That is why so many mainlanders envy them. Of their nature they break down the multiple alienations of industrial and suburban man. Some vision of Utopian belonging, of social blessedness, of an independence based on cooperation, haunts them all. Tresco is leased and managed by the Smith family, who have generally brought in outsiders to work there. I asked another native of Bryher what he thought of Tresco. He spat over the lee gunwale of his boat, which may seem ungrateful, in view of all the Smiths become Dorrien-Smiths have done, and are still doing, for the economy and conservation of the Scillies as a whole; but the spitting was, I knew, not against man, but against principle. He was prepared to make most of his summer living ferrying holiday-makers to the place; he allowed the charms of its modern hotel, its jolly pub, its famous subtropical gardens; but he would leave the Scillies sooner than live on Tresco himself. And he used finally a phrase that was almost one of pity, as if speaking of a fat girl trying to be a ballerina. "It's not island," he said.

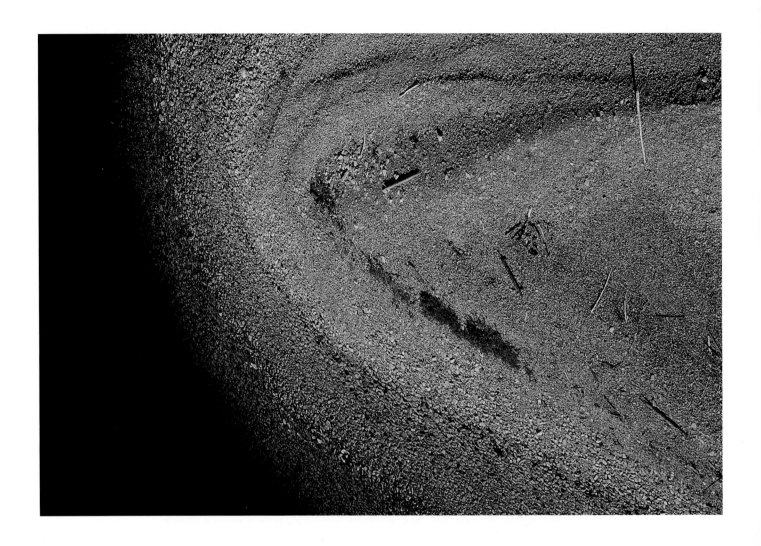

Since the proximity of the sea melts so much in us, the island is doubly liberating. It is this that explains why indigenous small-island communities, at least in the long-discovered temperate zones, are on the whole rather dour and puritanical in their social ways and codes. They have to protect themselves against the other perennial temptation of the island: to drop the necessary inhibitions of mainland society. Islands are also secret places, where the unconscious grows conscious, where possibilities mushroom, where imagination never rests. All isolation, as the cold bath merchants also knew, is erotic. Crusoes, unless their natures run that way, do not really hope for Man Fridays; and islands pour a stronger wine of forgetfulness of all that lies beyond the horizon than any other places. "Back there" becomes a dream, more a hypothesis than a reality; and many of its rituals and behaviours can seem very rapidly to be no more than devices to keep the hell of the stale, sealess, teeming suburb and city tolerable.

All desert islands, perhaps all desert places, are inherently erotic, as countless stranded individuals have realized. We all, whatever sex we are, want to know why we are alone; why has that universal yet obstinate human myth, the one that we generally call God, so forsaken us?

I have always thought of my own novels as islands, or as islanded. I remember being forcibly struck, on my very first visit to the Scillies, by the structural and emotional correspondences between visiting the different islands and any fictional text: the alternation of duller passages, "continuity" in the jargon of the cinema, and the separate island quality of other key events and confrontations — an insight, the notion of islands in the sea of story, that I could not forsake now even if I tried. This capacity to enisle is one I always look for in other novelists; or perhaps I should say that none I admire lacks it. It is a capacity that lies quite literally at the heart of what has often been called the first modern novel, Defoe's *Robinson Crusoe*; and it lies equally at the heart of the very first novel of all, Homer's *Odyssey*. This island remains where the magic (one's arrival at some truth or development one could not have logically predicted or expected) takes place; and it rises strangely, out of nothingness, out of the onward dogwatches, mere journeying transit.

The word isolation comes, through Italian and French, from the Latin for island, *insula*; from which we more directly derive insulation. Thus from the same mother-word we have both the active and the passive, the painful and the protected aspects of being cut off. The ancients rather prettily derived insula from *in salo*, "in the salt sea;" but it comes from a primordial and widespread root that also surfaces as the Greek *nesos* and the Irish *innis*.

Our own quite disconnected island comes from an Old English word that had no s — it crept in, unpronounced, only by scribal confusion with the medieval French isle. Yet there is a kind of justice in that surreptitious s. For s and l are great water-consonants, the sound of the sea and the sound of the sea on land, surging and washing, lashing and lapping; insula, isolation, soleness, solitude. "Soul," in the dark backward and abysm of linguistic time, is another universal European word, meaning transient, and sounds much more maritime and soughing in its Old English form, *sawol*, to which the Greek *aiolos*, fleeting, shifting, as changeable as the wind-god Aeolus, is connected. And what more natural than that ancient man should gain his sharpest sense of both physical and metaphysical loneliness ("all-oneness," nothing but one soul) at sea, and associate the concept to its sound — perhaps above all on the most formidable of the seas he knew, and named after Atlas?

I first began to feel the releasing power of *The Tempest* when I lived on Spetsai, my island in Greece — the lack of a Prospero, the need of a Prospero, the desire to play Daedalus. It is the first guidebook anyone should take who is to be an islander; or since we are all islanders of a kind, perhaps the first guidebook, at least to the self-inquiring. More and more we lose the ability to think as poets think, across frontiers and consecrated limits. More and more we think — or are brainwashed into thinking — in terms of verifiable facts, like money, time, personal pleasure, established knowledge. One reason I love islands so much is that of their nature they question such lack of imagination; that properly experienced, they make us stop and think a little: why am I here, what am I about, what is it all about, what has gone wrong?

Why do the fastest lobsterboats and the finest basketball players come from the same island?

SANDRA DINSMORE

BOATS AND HOOPS

"IF YOU KICK OVER a rock, you'll find a boatbuilder on Beals Island," goes an old saying. The boatbuilder adage applies to lobsterboat racers and basketball players as well. Beals Islanders have played on many state high school basketball championship teams, and while fiberglass has eliminated most of Beals' boatbuilding shops, champion lobsterboat racers and basketball players remain the norm for this small downeast community.

Beals, itself, is made up of two islands: Beals Island is about a mile long; Great Wass Island, about seven miles long, connects to Beals by a causeway. The majority of dwellings are concentrated on Beals, and concentrated is the right word: the houses at Mack Point, homemade for the most part, are jumbled together so closely that if you spit out the window, you might hit your neighbor. Property lines appear nonexistent, though everybody knows who owns what, or they don't care. Three active churches foster community closeness of another kind. In winter, piers loaded with lobster traps line the shore on either side of the Jonesport-Beals Bridge, announcing the islanders' main occupation, as do seasonally boat-choked coves.

That 600 people can produce so much success seems due to familial closeness, respect for each other, lives of hard work, stubbornness, determination and competitiveness; all of which breeds confidence.

Most Beals islanders are related, descended from one or two common ancestors. Manwarren (originally spelled Manwaring) Beal, whose family arrived in America in 1621, settled on what was then called Little Wass Island about 1765, changing its name to his; Captain John Alley arrived in 1770. Beals and Alleys still predominate, but other families have added diversity to the island.

Basketball player Sandi Carver, 23, grew up in a gym along with teammate Jan Beal, who's now playing basketball at the University of New Hampshire. After playing high school basketball, Sandi played at the University of Maine for four years and loved the support she got from townspeople back home. In Orono, she feels, parents come to games, but at home, she says, "It's not just your family, it's all of your family and all of your friends and everyone in the two towns; everyone comes out to watch the games." She thinks it's because everyone knows and cares about everyone else and, because there are no movie theaters or shopping malls, it's entertainment. Beals Islanders get right into it, yelling suggestions and comments, riding the referees, loving every second.

It's the way they do everything — full tilt. Take the way generations learned to build boats. As youngsters, they visited around the island's many boat shops — at one time 13 crowded its shores — looking, listening, analyzing; then they went home and designed and built their own toy or play fishing boats or had one made by a boatbuilder. They played at fishing with those boats until they became fishermen and built their own full-sized ones.

Facing page (top): Boatbuilder Ernest Libby Jr., 63, started making models at age 10. Below: Basketball comes as naturally to Beals Islanders as boatbuilding.

Above: Bennie Beal's 46-year-old STELLA ANN has roared past the competition to hang onto the record for the world's fastest lobsterboat.

"If I wanted a house, I built a house; if I wanted a boat, I built a boat."

Dwight Carver (left photo) played on Beals Island's championship teams of 1970-73. His daughter Sandi (right photo) starred at the University of Maine in the 1990s.

Isaac Beal, 57, who now manages a salmon farm, built boats for 19 years. His living room is a repository of Beals boat designs and of two- and three-masted play boats built by his grandfather and uncle. He says, "I had a fella 90-years-old tell me [when he was a boy] he waded and walked clear to Alley's Bay [on Great Wass Island], bought a play boat off'n my Uncle Floyd, and sailed it two miles back to his house." As a youngster, Isaac himself had a lot of play boats and swapped them around the way city boys swapped baseball cards.

Boats always fascinated boatbuilder Ernest Libby, Jr., 63. "I was designing boats at 10," he recalls. "I always made toy boats — play boats; tow 'em around the cove on a string." He'd go around the boat shops to see how the boats were built, and as soon as he could work a plane, hatchet and a few other tools, he began making two-foot-long sailboats and lobsterboats. Libby played with toy boats all summer long, making and setting out small trap buoys along the shore. His sons did the same thing, then the practice stopped. Last summer, though, he saw play fishing boats in the coves, so it's coming back.

Earl Faulkingham, 48, remembers having Libby build him a play boat, which he still owns. Then 10 or 11 years old, he paid for the boat by digging clams at $4 a bushel and says, "I took real good care of it, it meant so much to me: as much as a car to a teenager today." Earl and his friends played at fishing until they went to high school.

High school in Earl's day meant attending Beals High School for the last year before consolidation, for lack of students, with Jonesport High School in 1968. Years before, students from Beals had to cross Mooseabec Reach by boat to attend Jonesport High. Many parents worried about their children making the trip in rough weather, they decided to separate from Jonesport and have their own high school. Three times a group from Beals went to Augusta to plead their case. In the end, their stubbornness paid off with the incorporation of Beals Island, in 1923, as a separate town. Islanders, not the state, built an addition to the elementary school for the high school, which graduated its first class in 1925.

Its students excelled in many ways: Norma Wilcox (now Backman) brought home the first trophy for the Washington County Public Speaking Contest in 1942; it returned again in 1952, 1955, 1957 and 1958.

Beals basketball triumphs typify the islanders' determination to succeed: in the 1950s, Beals High School won three state championships despite a home court of only 50 by 25 feet and hoops only 40 inches from the low ceiling. In 1950 the entire school had 32 students, 17 of whom were boys. Eleven of those 17 made up the team. And they won. As Junior Backman (Herman Backman, Jr.) puts it, "Whatever boys there were, were on the team." In the 1970s the Jonesport-Beals team won state championships for five years straight and over the years brought home 19 gold balls.

Dwight Carver played on the championship teams of 1970-1973. He speaks of the members having "a tremendous love and respect for each other," and says that every Sunday morning he and another team member visit a third who now lives in Milbridge. "When we grew up," he explains, "everything we did, we did together from the time we were big enough to start running around the dooryard till we were 18."

By 18, most Beals boys go fishing for real. One kind of fishing predominates, lobstering, and one kind of boat, the Beals type, is designed for it, made by fishermen for fishermen.

Lobstering in the extreme waters of downeast Maine has always been difficult. Fishermen must haul their traps in tune with the tides, which rise and fall some 30 feet — the height of a three-story building. Tides run so hard out in what they call the "main tide," away from the shore and islands, fishermen usually haul when the tides are slack: at what they call low water slack and high water slack. When the tide is running at full speed, it can make eight to 10 knots. Fishermen sometimes have to wait hours for the tide to slacken, especially when it's going out. At ebb tide, the force of the water drags buoys under.

THE PRACTICAL-ESTHETIC COMBINATION
Bill MacDonald, marine resources director at the Island Institute in Rockland, used to own a cedar-on-oak Osmond Beal 37. He says it moved very well through water, yet had tremendous stability and comfort. Although boats from other places may be made for utility alone, MacDonald thinks Beals boats combine practicality and usefulness with aesthetics. "They have a certain feel and mystique," he says, "you can't beat it."

Beals' engine-powered lobsterboat building started early in this century. Vernal Woodward, 91, thinks Morris Dow and Alton Rogers built torpedo-stern boats at Beals before William Frost arrived from Canada, though Frost generally gets the credit. Harold Gower and Riley, Alvin, Vinal and Floyd Beal worked with Frost, then continued on their own. George Brown worked alone. Today's boats are based on Frost's and Gower's designs.

Because Beals boats are wider than most, they have a more stable fishing platform and can hold more gear. Wider now than they used to be, Libby's 34 has a beam of 13 feet. Those made a few decades ago were about three feet narrower, and the old torpedo-stern boats weren't more than seven feet wide.

Within the general Beals Island style, designs by different builders are identifiable. "You can tell one of Calvin's by his windshield," says Osmond, 67. (It slants forward instead of back.) He adds, "Willis always puts on a longer cabin than we do," and says the bilge on his boats is rounder than the V-shape found on the other local craft. Three years ago Isaac Beal spotted an Ernest Libby, Jr., boat with its high bilges, in an Alaskan boatyard and says, "Of course, I recognized it in a minute." Libby's boats, noted for their speed, are sought-after by lobsterboat racers.

Beals boats enjoy a legion of admirers. "They have a certain feel and mystique," declares one satisfied customer.

The races originated on Mooseabec Reach well over a hundred years ago, back when lobsterboats carried sails, not engines. Those yearly races have done nothing to dampen the competition between Beals Islanders and anybody else. Dana Rice, of Winter Harbor, recalls, "Three days before the race there was no limit to the things they'd do to beat each other, legal or illegal." They'd plane down their boats, even break the glass on their windshields during a race to lower wind resistance. Bennie Beal, though, is the ultimate competitor: his 46-year-old oak-and-cedar STELLA ANN has roared past all younger, lighter opposition to hang onto the record for the world's fastest lobsterboat.

Today only a few boatbuilders remain on Beals: Calvin and Osmond Beal, and Ernest Libby, Jr., build boats only in winter, and they're all fiberglass. Five different boat shops build Calvin's designs, H & H Marine in Steuben builds Osmond's and Young Brothers in Corea builds Libby's. Willis Beal, who mainly designs and builds wooden boats, works on and off. R P Boats in Steuben builds his fiberglass designs.

Boatbuilders changed over to fiberglass because upkeep on a fiberglass hull is minimal; because a wooden hull calls for 1,000 hours of labor and one of fiberglass, 40; and because good oak and cedar were harder to come by.

In the days before there was a car ferry from the mainland to Beals, Libby says they'd have to float the boat lumber across Mooseabec Reach, the oak tied on top of the cedar so it wouldn't sink, then they'd haul it up the bank to the boat shops. In 1958, a 1,000-foot toll bridge joined Beals to Jonesport, bringing with it accessibility; animals hitherto not seen on the island (skunks, raccoons and foxes); and increased intermarriage between Jonesporters and Beals islanders. Some people thought the bridge would be the ruination of the island, but they were wrong. Forty years later, the summer visitor has crossed the bridge and taken up residence; but, as with the skunks and coons, Beals Islanders have learned to tolerate this new animal, too.

Sandra Dinsmore *writes regularly for Island Institute publications.*

A fisherman must follow the routes herring travel in order to build a weir that will catch them on rising and falling tides.

RUNNING

Aquaculture transforms an island community

INSIDE THE BLACK PRINCE WEIR, nine men with their skiffs tied together sit captive to a story being told by the oldest among them. Animating his story with hand motions and island lingo, the senior man's is the only audible voice. As he delivers the punchline, a burst of "ho hos" and "ha has" bounces off the cliff directly behind the weir. Cigarette butts get tossed into the water; three men move their skiffs and hop into the boat closest to the net inside this giant herring trap. Together they pull hard on the net, the tiny skiff heaving and rocking as the net slowly surfaces, encircling a foaming pool of thrashing fish. Bringing their skiffs around close, the rest of the men dip their nets into the shrinking pool, pulling out the wiggling silver fish they will use for lobster bait. They dump them into bright green plastic baskets. As the net gets pulled higher and higher, the sound of the herring splashing is rivaled only by the seal-scarer from a nearby salmon farm.

Less than a mile away, salmon leap out of the water, mouths open, as young men spray pellet feed into their underwater cages. Since its introduction to Deer Island nearly 20 years ago, aquaculture has presented this depressed New Brunswick community, just off Eastport, Maine, with a multi-billion dollar industry. But with the promise of economic prosperity came a threat to a traditional lifestyle dominated by herring and lobster fishing for over 150 years. Aquaculture has forced many to take sides, and islanders continue to face the challenge of change as they struggle to maintain the fabric of their community.

"We keep going"

The sun rises orange over Campobello Island, starting a day begun hours earlier by Dale Mitchell. The sunrise tips and disappears as the FAMILY PROVIDER, his boat, heaves in the swells. Two small trees he has cut off his property to use for repairing the weir trail behind. With Mitchell is his brother-in-law, Gary, who co-owns the Abnaki weir. This morning they will start preparing it for summer, the season for catching herring.

Dale Mitchell has been working these waters for most of his 41 years. "It's all I ever did," he says. "It didn't occur to me to be a noble thing. But my father bred it right in me. And I love to do it. I figure I'm successful at it. It's all I ever knew. It's me."

Before building the weir in 1986, Mitchell spent four years watching the herring move around the tiny islet where it is located. Knowledge of an area is imperative: a fisherman must follow the routes herring travel in order to build a weir that will catch them on rising and falling tides.

Mitchell points out an eagle's nest. "That one's been there about 10 years now," he says. "They get bigger every year. Some eagles' nests have been around as long as my father can remember."

Weirs may be owned by as many as nine or ten people, often within extended families. "It just kind of works out that way," he explains. This tradition of joint ownership has kept the community tight over the years, as the work of one becomes the work of many. "Deer Island is a place where, if people see that you'll work, they'll make sure you get by."

Aquaculture, introduced to Deer Island 20 years ago, has presented islanders with new opportunities but forced them to take sides.

TOGETHER

K.J. VAUX

Behind each weir is the steadfast tradition of its name - for location (Grass Point), for the men who built it (The Bachelor), for historical people or tribes (Abnaki) or events of the year they were built (the Jubilee weir, constructed in the 50th year of Queen Victoria's reign). The Zig Zag weir was named for its shape; the Black Prince weir refers to the color of the water when the weir is thick with trapped fish.

Dale Mitchell's father, George, and his son, Judson, pull up in a skiff and join in the work. George isn't as active as he once was, but he is still out on the water as much as he can be. "I'm 73 but I don't quit yet. I know the day I do, just sit in the house, that'd be the end." Judson spends his summers on the water with his father, learning to build and tend weir. At eight, he is proficient with his hand on the outboard motor, and takes himself and his grandfather to the scow where Dale and Gary are "building" the weir: lining it with synthetic twine nets and replacing and mending broken poles. Judson hops on the scow behind his grandfather, as three generations of Mitchell men work to prepare their weir.

"If I had a bad year last year, it wouldn't even occur to me not to get the weir ready for this year," Dale says on the way back to the wharf. "We keep going. It's not always an every year thing, but we keep going. It's like the deer hunter. You may go out, you may not get one every day, but one day you will."

"Long and steady wins the race," adds his son.

"Aquaculture has been good to us"

From the turn of the century through the Second World War, the waters of Passamaquoddy Bay were filled with herring. Weirs could be seen from every water view of the island, and traditional fishing was as reliable a source of income as it was a rubric for island culture. The 1950s, though, saw a drop in fish stocks and prices, and slowly Deer Island fell into a depression, losing more than half of its one-time population of 2,000. Campobello Island, Grand Manan, Eastport and Lubec all suffered equally, and the population thinned as companies closed and workers and owners left.

With Connors Brothers still running a sardine plant in Black's Harbour, New Brunswick, some fishermen were able to eke out a living and continue the yearly cycle of traditional fisheries: trapping lobster in the spring and fall, scalloping in the winter, tending weir in the summer. Like Judson Mitchell, children started working at an early age, acquiring the environmental, mechanical and business knowledge of traditional fishing.

But even for fishermen who survived the hard years of the 1950s, the boom-and-bust cycle of the sardine industry since then has made it tough for many of them to practice the lessons their fathers taught them. While trapping and scalloping provided some income, many traditional fishermen became increasingly discouraged by years of empty weirs.

Photographs by Heather D. Hay

As the sardine industry declined, aquaculture provided jobs for islanders who otherwise would have had to leave.

At 7 a.m., Allison Pendleton motors through the dense white fog to his salmon farm a few kilometers from Lord's Cove. Allison has been on the water since he was 15 years old. "I left school one day, and went dragging scallops the next," he says with a smile, recalling the distracting view of harbor traffic from his school.

Allison farms 80,000 salmon in 32 floating cages. Each cage is supported by circular, hollow plastic tubes. Beneath them, nylon nets drop 36 feet, containing his salmon in a cone-shaped area where they are fed three times daily, sometimes four. Today his crew will harvest about 2,000 salmon. Although Pendleton was one of the first Deer Islanders to enter aquaculture 13 years ago, he was originally a traditional fishermen, tending weirs and trapping lobster for nearly 40 years. But like many in traditional fisheries, falling prices and diminishing stocks of herring made it harder and harder to earn a living. "Herring got so bad that we was going seven, eight years, and every year going in debt. I just couldn't go through another one. I woulda been completely broke," he recalls.

Recently introduced to the area, aquaculture offered an alternative. "Most of [the farms], like ours, we had a weir and it wasn't catching any herring so we sawed it down and put an aquaculture site in there." While it was a business with its share of risks, the profit margins looked promising, and his family was willing to go into the business with him.

Aquaculture presented Pendleton with the continued freedom of being his own boss, managing his own business and working on the water. It provided a steady income, without the ups and downs of traditional fishing. "Aquaculture has been good to us. We never get rich, but we make a good living, something we never had in fishing." While the hours are long, and the work is hard, "you know what you're doing. And you're free to do what you want."

Pulling up to the cage he will harvest, his crew brings the salmon to the surface by pulling the seagrass covered net almost entirely to the surface. Using a mechanized dip net, they scoop a few dozen fish out from the pool and dump them into a holding tank on deck. Two men grab fish and lay them flat, while a third makes a quick cut under the gill and shoves them towards a container of ice and sea water. Some glide listlessly, others kick and flip frantically. Once in the tanks, scarlet plumes rise to the top as they thrash and pump their own blood into the salty ice water. This method not only keeps them alive longer, but leaves less blood for workers to clean at the processing plant.

Cruising home past a dilapidated weir and the salmon farm directly next to it, Allison reflects on the advent of aquaculture on Deer Island. "It's not as bad now as when we started. When we first started it was hellish. People wouldn't speak to you. They wouldn't go to church. Wouldn't associate with you at all. But most of that's gone now because most people on the island are involved in aquaculture in one way or another."

•

Earl Carpenter, owner and manager of Deer Island Salmon, felt this tension when he first came to the island to establish his business. "In 1985 when we first came here, there was very strong resistance to doing anything differently. The people that were still here, quite rightly, had the notion that it would change their lives if something new came along that was dominant; and change is a hard thing to accept when you don't know what the outcome is going to be." Yet by stressing the importance of community, Carpenter feels he made important inroads among skeptics. "We had a very strong emphasis in the early years of putting money back into the community where we could," he says. "And the people sensed that here, and they went with that. They said it made sense."

What initially made sense to some islanders was the jobs aquaculture promised. The sardine industry declined during the 1980s, and in 1993 the sardine factory in Fairhaven closed, leaving nearly 20 percent of the island out of work. Aquaculture, meanwhile, provided jobs for islanders who otherwise would have had to abandon their homes, communities and communal history in search of work. In addition, the salmon industry provided a sense of future for young islanders.

Earl Carpenter looked closely at the island's workforce when he located his company here. "Most of the employees on the cages tend to be young single men because they can work at night, on weekends," he said. "Their time is at your disposal more, plus they have the physical attributes to do what they do out there. In our processing plant, 90 percent of the people are women. It suits them very nicely with the work hours they can make available. It becomes a way for family participation. And to have them both together in the community makes it easier for me to manage."

His processing plant is filled with both those left out of work by the collapse of herring, as well as those new to the workforce. Dressed in long yellow rubber smocks and clear blue aprons, eight women stand along a sterile metal trough, each working on a specific task of gutting and cleaning salmon. Beyond them, four young workers label, package, wrap and palletize salmon for the delivery trucks waiting outside. Upstairs in the office, three young women and two men manage the business, contacting buyers and distributors from New York to Montreal.

Farmers and fishers

For those who once packed sardines, aquaculture has meant not just employment, but better wages and far better working conditions. "I used to pack fish in Fairhaven," says Grace Regression, 58. "You stand there eight hours a day and you don't lift your head because the more you pack, the more you make. And that's slave labor, I'll tell you. And they come out of there five years later and their hands all twisted [with] arthritis. This is more relaxed. I get a chance to step outside for a puff," she continues, sharing a cigarette with her co-workers as they stand outside on one of their hourly breaks. Also working at this plant are young adults, too young to have packed fish in Fairhaven, but old enough to have looked for work off the island were it not for the jobs salmon has provided.

"Without salmon here, there wouldn't be any young people left on this island," says Chad Stuart, 26, who worked for five years on a salmon farm.

"People [are] getting all upset about the salmon sites being here, but if you want the people here too, that's how it'll be," Stuart says. "We could be doing the same thing on the mainland as we're doing here. Where would that leave the island? Even if you went somewhere else, where do you start, Burger King?"

The strong economy on the island has made it possible for Chad to build a home for his young family.

For Tyson Richardson, 25, his salaried job at his father's salmon farm is the foundation for a house he plans on building just up the road from his parents. While the majority of Tyson's time is devoted to salmon, working for his father has also meant

"A fisherman can do accounting, can do mechanical work, can do carpenter work. He can rig twine. Anything he'd need to do, he can do it."

helping tend and build their remaining weirs. Without salmon to keep him on the island, Tyson may not have learned the trade of his family in the generations before aquaculture.

•

Aquaculture brought the jobs that keep islanders, young and old, at home, but those jobs have also brought outsiders. Many come from nearby mainland communities to work. In particular, Deer Island has a growing population of Newfoundlanders - left jobless by the cod moratorium, closing paper mills and dwindling employment opportunities back home.

Working the forklift at Deer Island Salmon processing, Marie Sharpe chuckles that she has "the worst job in the plant" as she dumps containers filled with salmon, blood and ice down into the trough. But she is grateful for her job: "I would do any of it," she says. Like many unemployed Newfoundlanders, Marie came from her island to this one, knowing that both she and her husband could find steady work.

Many native islanders are uncomfortable with the growing population of people from away. "I don't know anybody on this island anymore," says an older resident of Leonardville. "They've all come for jobs, but when aquaculture leaves, this island's going to be in as bad shape as before. And we're gonna be the ones left with it. It's our home. Aquaculture was set up for the people of the island for when they had a bad year in the sardine business. Every year wasn't a good year. You knew that when you built weirs. This was the way of life."

•

The "way of life" for generations of island fishermen has been defined by the work ethic of traditional fisheries, stemming from the self-dependency, ingenuity and independence necessary for the wide variety of work traditional fishermen must do. "A fishermen can do accounting, can do mechanical work, can do carpenter work. He can rig twine. Anything he'd need to do, he can do it. He can educate himself to do it if he needed to do it," Dale Mitchell boasts. "The best part of fishing is doing good. I'd rather catch 200 pounds of lobster at $5 a pound than 100 pounds of lobster at $10 a pound. When you're through, you've got more sense of fulfillment, more satisfaction to see that boat loaded. You can say 'Boy, you had a good haul!' Aquaculture brought a new work ethic to the island. "Every other Thursday that check is waiting for me," says Bradley Hurley, who works on one of the salmon sites.

Traditional fishermen have been reliant on a precarious, often diminishing, natural resource. Aquaculture, on the other hand, is internally-dependent and industry-driven. The difference between farmers and fishers has sparked conflict on the island as many islanders, weary of an uncertain financial future, were attracted to the stability they saw in aquaculture.

"It's a steady thing you can always count on," says Chad Stuart. "With weir fishing or trapping, you never know if you're going to have a good year or not. That's just the way it is."

While these hourly wage jobs may provide more stability than traditional fishing, Dale Mitchell worries about the effects on island life and community. "To give someone an incentive to share in the profit, I think it makes a difference. People are much more up front, they keep things up, and they keep things going," he says. "But (in aquaculture) they don't see that the small things they do help the profit at the end of the year, because they don't share in the profit. It's just a job, and it's a boring job, so they're not gonna put more in than they got to. It will become just another place where people work by the hour. I think that ruins a community, because nobody has any pride in what they do. It's just a job. It's not a way of life. The whole way of life is being lost, because people have lost their sense of what it takes to be successful."

•

It is from the personal demands and challenges of working in traditional fisheries that many fishermen draw a sense of pride and independence, a key to keeping the island community together. And with these personal demands for the most part absent from aquaculture jobs, some feel the cultural price of salmon is too high. As more and more islanders work on sites and in processing plants, many fishermen resist the factory-town mentality cast on the island by the salmon industry. The resistance is directed, at times, toward the non-islanders who brought

Aquaculture has transformed fish processing as well as fish-catching on Deer Island. For those who once packed sardines, aquaculture has meant not just employment, but better wages and far better working conditions. At Deer Island Salmon (right photo) each worker in the production line performs a specific task.

aquaculture to the island, and who continue to control most of it.

Even nuances in island language have come to separate fishing from aquaculture. For traditional fishermen, a "scow" is motorless and used while attached to a larger boat or skiff. What salmon farmers call a "scow" is longer, has an engine and a small heated room, and is the main working vessel for feeding and harvesting. To them, what fishermen call a "scow" is actually a "barge."

Despite disagreements over the communal effect of aquaculture and changes it has introduced to the island, the wealth it has brought is readily apparent. New cars and trucks crowd the island's six wharves. Property values have jumped, and island villages are now dominated by well-kept homes, many of them over a hundred years old, handsomely refurbished and maintained. For most in the salmon industry, the improvements to their daily lives outweigh what many consider to be necessary losses.

Perils of prosperity

While aquaculture has presented change to life on the island, so has it challenged life in the space that surrounds the island. The presence of salmon in the waters originally monopolized by traditional fisheries has raised voices and opinions across the island as farmers and fishers attempt to coexist within the same fragile and limited sea space. Many traditional fishermen feel aquaculture jeopardizes the integrity of the environment on which they rely. In particular is the competition for space, and the location of salmon sites, often placed directly in the "fish way," the routes herring travel that lead them into weirs.

"There isn't room for both." says Stanford Young, a retired weir fisherman. "You put them cages in, and a lot of that trash through the water, and the feed and such, and herring won't come."

Since the introduction of aquaculture, Gary Conley has harvested only half as many herring. Donny Richardson went from 11 working weirs to just three, only one of which catches any fish. In some situations, salmon farms have been located within 1,500 feet of a weir, rendering the wooden structure useless.

While some salmon farmers acknowledge that site locations can affect the flow of herring into weirs, they do not readily accept responsibility for the decline of herring populations, or for empty weirs. Many of these once traditional fishermen are quick to point to the decline of traditional fisheries long before aquaculture - why aquaculture was targeted for Deer Island, and why so many traditional fishermen switched.

More hotly disputed by traditional fishermen is the seafloor beneath salmon cages, often thick with fish excrement. Although the government requires this build-up to be photographed and monitored, the broader effects it may have on the ecosystem worry many in traditional fisheries. While the 28-foot tides often carry much of the refuse away, precious lobster bottom is threatened by the salmon above it.

"[Stoltz Sea Farms] planned on putting a site right on top of two weirs that was catching herring. My god, there's no better place in the whole area for catching lobsters. I don't blame the fishermen for howling about that. They got right into turmoil," says Allison Pendleton. This environmental concern has slowly turned political, as fishermen felt the need to protect their interests. Pendleton tells the story of when surveyors set buoys for the planned site, only to have them mysteriously cut and washed ashore the next day. The site was never put in.

Such recourse is not endorsed or discussed, but advocates of traditional fisheries have been left with few alternatives. During the first 10 years, requests for salmon sites were automatically filled without advertisement, leaving fishermen without political power to stop the expansion of aquaculture. While written protestations to the government are now available to those opposing the creation and sale of salmon farms, public hearings are still unknown to island fishermen.

Dale Mitchell, one time president of the Fundy North Fisherman's Association, also believes a lack of education on the part of the government has led to the neglect of traditional fishing rights. "We've been here since 1840 to go herring fishing. We should have had some kind of priority, with that tradition behind us, to have some respect for what we did."

Mitchell feels the government has been swayed by the money invested and created by aquaculture, and has failed to negotiate a compromise between the two industries.

"You gotta keep up with the times"

Despite the fractions and deep disagreements that aquaculture has incited on the island, it has coexisted with traditional fishing for nearly 20 years. In that time, islanders have become more comfortable with the industry, and the change it has brought. Even the staunchest traditional fishermen have come to accept that aquaculture is on the island to stay, and that a majority of islanders have benefited, at least economically, from it. Unlike Campobello and Grand Manan, which have sizable tourist industries, Deer Island has retained its identity as a working island. And as aquaculture and traditional fishing innovate and advance, so will the island and its people. Aboard the FAMILY PROVIDER, Dale Mitchell scans a small screen that flashes information from his Global Positioning System. "I didn't want to get it, but you gotta keep up with the times," he says. Dale and his wife often spend hours on the Internet, gleaning information they can find to help him keep up with the elements, other fishermen and the salmon industry. Recently introduced to salmon farms on the island are automatic feeders, which change the nature and numbers of work on marine sites. And as fish health and management have come under closer government scrutiny, the battles fought by both industries stand to become increasingly political. Changes will push both industries to innovate and evolve. And in the process, islanders too will also be shaping their identity, one as much born of the past as created by the present.

•

As the sun sinks behind the mainland, an opaque moon rises on the islands. The soft orange light that marked the dawn falls low on the water, and a boat blades across the reflecting surface where Abnaki legend claims the god Glooscap turned a pack of wolves chasing three deer and a moose into islands. From the air, the moose (Eastport's Moose Island) leads the three deer (Grand Manan, Campobello and Deer Island) as the Wolves follow in close pursuit, the islets among them dirt kicked up by the deer in their dash for survival. Like their namesake, Deer Islanders continue to keep one step ahead of the forces that threaten their survival. And although initially with trepidation, they have run together with the challenge and opportunity of change within the waters of the world's largest tides.

A freelance writer, K.J. Vaux is at work on a documentary film about traditional communities.

HARD TIMES GOOD TIMES AND THE SEAT OF THE PANTS

Nine decades on Chebeague

DONNA DAMON

Ellsworth and Melba Hamilton Miller were born on Great Chebeague Island, in the early years of the 20th century. Their birthplaces faced the open ocean and the rising sun. Surrounded by extended families with deep island roots, the Millers' childhoods differed greatly from those of their four grandchildren who are growing up on Chebeague today. Their lives have revolved around their meeting basic needs in harmony with the rhythms of the sea.

Ellsworth and Melba never met until they were teenagers, and then it was nearly 25 years before they married. During those 25 years Melba married, divorced and took care of her children, Warren and Alice Doughty. Ellsworth stayed at home and helped to support his parents by fishing, clamming, scalloping, sardining, trawling and lobstering. During World War II he sailed aboard merchant vessels carrying men and supplies across the Atlantic.

Ellsworth Miller is 88 today. Melba Hamilton Miller is 86. They were married in 1949 and raised their children, Warren, Alice, Donna, David and a granddaughter, Carla, on Chebeague. Two of those children, David Miller, a lobsterman, and Donna Miller Damon, who conducted this interview, live with their families on the island.

Who influenced you when you were children?

ELLSWORTH:
My biggest influence was my mother and her strong right arm and her big hand. It just fit the seat of my pants. That's the way she did it, either that or a switch stick. I used to chase my father around all of the time. When I was four years old, he went halibuting. I used to go out with him when he set his halibut gear. He had a 21-foot Hampton boat with a Hartford engine, a make-and-break. He'd go off the Cod Ledge. We'd leave home about 11 o'clock in the morning, and we'd get off there and those bait seiners would be seining those big sea herring and mackerel. We'd get a barrel of bait and he'd cut those fish right in two. He taught me about the fishing grounds.

MELBA:
My mother and father influenced me. They taught me about religion, church and God. They taught me about being honest and the Ten Commandments and all that. My father loved to go to church and I went to Sunday school every Sunday. My folks gave me good advice and they were good people. I tried to be like them. They brought me up religiously. Living around family was part of living on Chebeague.

Above: Ellsworth and Melba Miller at their home on Chebeague in 1998. Opposite page: Ellsworth and Melba during World War II.

What did you do after school and summers?

ELLSWORTH:
When I was little I usually ran away and went down to the shore. If the men were down at the shore baiting up and stuff like that, we'd go down and play around in the punts. I don't remember not knowing how to row. When I was a teenager, I worked mostly. I cut wood, we had gardens and I went clamming. We used to go out and tease the neighbors. We'd put a devil's fiddle on a house. You take a spool of Aunt Lydia's thread, tie it on a fine finish nail or a horseshoe nail and stick the nail in under the window. Then run a piece of rosin over the thread. It would make an awful whistling noise. We played a little baseball, but not very much.

MELBA:
We played on the edge of the woods up above my mother's there. We had playhouses made out of fir boughs. We hunted around and found old teakettles and put them in our playhouses to see who could have the best playhouse. We used to all go up there and meet and, of course, there would be some fighting and pulling hair. I didn't do it but one or two did. We used to go to the shore, and take a picnic. Sometimes my father would take us up the bay in his boat, the MELBA. My brother, Louis, bought it for my father.

How do you think things have changed on Chebeague since you were young?

MELBA:
Well, you knew everybody here. Everybody. Now you don't know anybody. We called everybody aunt and uncle. We had more respect. We didn't call Chebeaguers "Mr." and "Mrs."

ELLSWORTH:
They had a community club here on the island. They had baked bean suppers and harvest suppers like boiled dinner. They charged 35 cents for a bean supper for adults and a quarter for kids and that included the entertainment. Some people could dance and other people could sing, some could rave and swear if you could get them mad enough. Anybody who could do anything to entertain, did it. There was a lot of community spirit. Some of the summer people would do that, too.

Pictured left to right is Winona Hamilton (cousin), Melba Hamilton, Vera Hamilton (sister), Louis Hamilton (brother), and Ruth O'Sgood (niece), circa 1918.

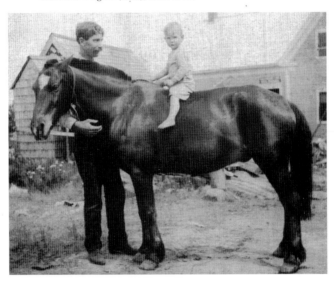
Ellsworth with father Ellsworth Haridon Miller, circa 1912

I can remember of my father going out in his little boat with Walter Calder. It was way below zero. I can see them now going out between [Chebeague and] Bangs, disappearing in the vapor. I thought they'd probably never ever come back. It was lucky they did I guess. They were fishing off around the Elbow, about three or four miles outside of Half Way Rock light.

Often as not it would be a blizzard when they'd come in. My mother would tell my brother, Leland, and I to go down on the wharf. She would say to Leland, who was a few years older, "You take Melba's hand and don't leave go of it and you ring the bell down there on the freight shed." She thought it would help them get in. Sometimes they said they heard it when they rounded Bangs over by Sand Island. In those bad winters, you had to go that way if you went at all.

How did the Great Depression affect the island?

ELLSWORTH:
The Depression hit this place just like everywhere else. When your source of revenue dried up, you picked up a dollar any way you could.

MELBA:
In the winter, [the men] shoveled snow. We had a lot of hard winters. I can see my father and Uncle Ad now. They didn't have to ask or anything, they just went down with their shovels. It was customary to keep the roads clear. It was expected of 'em, if a person was able to do it. Keeping the roads clear was an emergency. They shoveled until it was clear enough for Archie to get through with the mail and George Leonard could get through with the groceries. They got 35 cents an hour, but that 35 cents an hour would buy 10 dollars worth of groceries in today's money. The kids today don't know what snow is.

Did the Depression affect your life here on Chebeague as much as it would if you had lived in a city or someplace else?

ELLSWORTH:
We used to plant big gardens. We did it more during the Depression, so we'd have plenty of food that way. If I lived in a city, I probably wouldn't have had a job. Here we'd do anything we could get that they'd pay you for. There was always a market for our fish. The price weren't very high, but I've sold haddock for two cents a pound at Burnham & Morrill [the cannery in Portland]. They made fish cakes and fish chowder in a can, you know.

We had all the vegetables we wanted. We had a whole cellar full. Rachel [Ellsworth's sister, who was a student at Boston University] came home one year for Christmas. My father went down cellar to get some potatoes. She looked down there, and she saw here was six barrels of potatoes in the cellar he'd raised. She got to telling about the starvation in Boston. She was working in a settlement house in the North End of Boston. Those people were going up around Beacon Hill, robbing grocery carts that delivered groceries around to well-to-do people. Nobody around here had to steal for food. Sometimes we'd go fishing and bring home half a trip to give to the neighbors. The neighbors took care of each other.

ELLSWORTH:
People lived like the old people had to live. One time my father and I was fishing and we didn't have any turkey for Thanksgiving Dinner, but we had plenty of vegetables and plenty of pie and everything like that. So the day before Thanksgiving we went down to shore and borrowed an old scallop drag. I don't know who it belonged to, but we thought we'd take it and get a mess of scallops for Thanksgiving dinner. We went over by the Horse Island bell buoy and we got scallops enough to shell out a quart and a half. We had our trawls all baited and two of them hadn't been set the trip before. So I suggested we set those two trawl and get the rotten bait off from them, then Bill and I could go down to the shore on Thanksgiving Day and bait them up again. We set those and fooled around with that old scallop drag until it all came apart on us. We had scallops enough for a couple of messes. Then we went to haul those trawls and there was a great big cod fish, a big white-bellied codfish! There was our Thanksgiving dinner, I thought. We used to have them baked just like a big cusk. Before we got the trawls up, we had more fish than we wanted to eat for ourselves. My father said, "What are we gonna do with them fish that we don't want to take home?" I said, "Let's go to Portland and sell the fish." We got enough out of it to buy us a Thanksgiving dinner. My father went up to Diamon's Meat market up on Market Street. He bought a loin of pork, 11 cents a pound, and it weighed 14 pounds. We cut a roast off from each end of it, and we cut some pork chops out of the middle. We got that and we got a quart of cranberries, 10 gallons of gasoline, and $5 to bring home from what few fish we got there. And we still had the big haddock that we took home to have baked fish. We had food enough to last us two weeks, and we got our Thanksgiving dinner!

Melba, circa 1929.

Was credit issued at the stores?

ELLSWORTH:
There were two grocery stores most of the time and a lot of credit was given. George Leonard let people cut ice. The only ones he'd give jobs working on the ice were people who owed him. Then he would just let it go on their credit or they could take it up in groceries. He used Charleson Ice Pond. He wouldn't hire me because he would only hire two to a family.

How did the island change during World War II?

ELLSWORTH:
The place was infested with a bunch of soldiers. Anyone from here who had any gumption went to work. Most of our young fellas went overseas. They called it patriotism.

What was the island like during the war when the men were gone?

MELBA:
Oh, there were soldiers around here enough to keep the girls happy! The girls were all marrying soldiers. It was hard. You couldn't have anything you wanted because of rationing. It was tough to be a single parent here during the war. That was before I got in with your father. I lived with my mother and got some money from the state. I cleaned houses and picked crabmeat until one o'clock in the morning. I would pick out 10 or 12 pounds and go up on the early boat and sell it to the Columbia Hotel where Bill Brise worked. When they'd give me the can back that I brought the crabmeat in, there'd be butter and sugar and all kinds of things. Those things were rationed and I couldn't get much of them for my kids. They were very good to me there.

How did you and your children spend vacations?

ELLSWORTH:
I didn't take a vacation. I just kept on working.

You don't think we had vacations? We had the biggest vacation of all. I was telling somebody the other day how we would go down to the shore the day after

school closed and spend the summer in the camp. We'd go to Stave Island in an old dory to get driftwood to burn in that old enamel stove. We'd row to Crow Island jigging for mackerel at sunset and stop and get a bucket full of herring from the Trident that had just come out of the seine. We'd sit on the ledges eating our cereal as the sun came up. Now, don't you think that was a vacation?

ELLSWORTH:
It wasn't a vacation. We was just having a good time and had plenty to eat.

What were some of your toughest times?

MELBA:
I don't know. One time we were so bad off I wanted your father to go over and see Roy Hill [the island selectman] and get some help from the town. Do you remember, Ellsworth?

ELLSWORTH:
I was outside lobstering that winter and I got sick. We got by all right. We ate an awful lot of lobsters that winter! The only time I could go was when the weather was decent, when it was fit for me to get out there. I'd go clamming, so we had clams on hand all winter. Another time, I cut my fingers nearly off in the lawn mower. They were just hanging. Old Dr. Petterson sewed 'em back on. He didn't give me anything. It was May and I had to go lobstering, so I wore a rubber mitten on that hand. I had it all bandaged up. I went alone. Warren was in the Navy, and I didn't take a sternman.

Do you ever wish that you had had a driver's license?

ELLSWORTH:
I could go around the Portland waterfront and do everything I wanted to do. You couldn't do that today. There is no place to leave your boat and you can't buy anything along the Portland waterfront. I gave up on the waterfront when I stopped going lobstering. I'd go to Norton's on Long Island and if you needed anything that they didn't have, like a pair of boots or anything, they'd order it for you. I sold my lobsters there because there was no place in the docks where you could leave your boat. That was the late 1980s. Then I had a stroke

The author, between her parents, circa 1954.

and stopped going hauling altogether.

Mama, you raised children on Chebeague from 1935 to 1977. What was the biggest challenge you faced?

MELBA:
Trying to teach them to do right. They always went to Sunday school, and that helped a little bit. Now they don't know what Sunday school is. You didn't always have all the clothes you wanted so you had to take care of their good clothes, so they'd have something to wear.

Was it easier for you when your children went to school on the island?

ELLSWORTH:
The older kids didn't have much for school activities here. They went to school to learn and come home. A lot of kids then quit in eighth grade because they'd rather have what money they could earn than sit around and study ancient history. Now most of the kids finish high school.

MELBA:
There wasn't as much to worry about when they went to school here as there was with you younger kids when you went in that boat, traveling to the mainland. I worried about you getting stuck in the ice I don't know how many times, but one I can think of that I'll never forget. I can see your father calling Ern Ross and different ones, and Alger Burgess was calling here. They were all talking about what they were going to do. I said, "Is this serious or what? You act awful funny." I was up looking out the upstairs window and I could see the boat stuck in the ice. Then Daddy said, "If that boat gets between two big ones [ice cakes], it will crumple up just like a piece of paper." That's why I was concerned. If they got between the heavy ice and the shore, it would crush 'em.

What was it like to be the wife of a fisherman?

MELBA:
When times got hard we had lobster, crabs, fried clams and clam chowder, and another clam chowder and fish chowder. We used to have mussel stew, and steamed

mussels. I never had mussels in my life until I got in with your father. My mother wouldn't eat them. I tried to tell her that they were good and finally I got her one day to take a taste of the juice. I told her it tasted kind of like oyster stew. She said, "Yes, it does taste kind of sweet."

My father always used them for bait you know. When I was a little girl I used to take his dinner down to that fish house. There would be all of those great big trawl tubs with the rope and gangings all curled up with those orange mussels shelled out on the hooks laying in amongst the trawl and I'd think, yuck! My land, I never thought I'd be eating them.

Ellsworth with two grandchildren.

Have either of you ever regretted spending most of your lives on Chebeague?

ELLSWORTH:
Why should we? We had everything we needed.

How do you think this island has changed over this century?

MELBA:
It's not Chebeague anymore, if you ask me. There's so many summer people now that I hardly know anybody. When we were young there were some summer complaints, a few, but they went come Labor Day, which was it. Then Chebeague was Chebeague again.

ELLSWORTH:
Once the cottages were closed up, we had Chebeague to ourselves for the rest of the year. A few of 'em stayed and retired but most of them, the men, had to go to work. When the Depression came some of the summer people went home and never came back. Some of them lost everything they tried to make other people think that they had. During the Depression the summer people lived just the same as the island people.

There were a lot of these people who were supposed to have a lot of money, who left here and people never knew what became of them. They left and never came back. The business people left, but the college professors stayed, and their families still come here and kind of grew up in the place. They have a different feeling about us than some of the others did.

There was that fella Scott, that John Seabury ran the yacht for in Boston. He was a stockbroker in Boston. The last anyone ever saw of him saw him was in New York down on the Bowery. He had a sign on his back, "Eat at Joe's." When asked what he was doing, old Scott said, "My wife and my family all kicked me out. They are blaming me for all their troubles. I'm working for a dish of beans."

If you could tell your grandchildren something you'd like them to know about what it was like to spend most of the 20th century on Great Chebeague Island, what would it be?

MELBA:
I'd want them to know about hanging May baskets, and about the time I was sliding down the hill over there at Central and I almost slid right off the wharf. They should know that I walked to Cliff Island over the ice with my family to visit my sister Statie in 1918! I'd want them to know that they don't have the snow now like we used to have.

ELLSWORTH:
Well, I want them to just live their life the way they should. The kids are happy here. If the life a person lives makes them happy, well, it's all right. It doesn't matter what you do or where you live, the most important thing is that you are happy. My father taught me that. We just lived together and worked together. He was a pretty easygoing person who just liked to do the best he could under the circumstances. I tried to do the same with my children.

Up in the office of the old Central Wharf Cold Storage in Portland where we used to go buy trawl bait, there was a motto on the wall, "Do the best you can with what you have to do with." You couldn't catch haddock and codfish with a clam hoe, but you could get a mess of clams with it. I guess you could say that is how I've lived my life.

A Malaga Island Family and Home

Maine Historic Preservation Commission (3)

EVICTED

HOW THE STATE OF MAINE DESTROYED A "DIFFERENT" ISLAND COMMUNITY

DEBORAH DUBRULE

But enough of the poor colored race, we doubt not many of them have white hearts; and placed on this isle in Casco Bay for some purpose, though at present an unknown one. ... But such a spot of natural beauty as could here be made for a few summer homes ... and could this gem of an isle be depopulated and rebuilt what a change and what an imposing entrance to our beautiful New Meadows River.

- LAURIS PERCY, *CASCO BAY BREEZE*, AUG. 24, 1905

IN 1905, MAINE annexed Malaga Island, telling its 45 residents they would be cared for and protected. Seven years later, the state sentenced this same island community to death, evicting its members and forcing some into institutions where they would die.

Malaga's white, black and mixed-race families stood "convicted" of intermarriage, illiteracy and poverty, among other social sins. Their "trial," however, hadn't taken place in a court of law; it had occurred in the columns of newspapers and the minds of racist opinion-makers.

Shunned by people on the main, only three island families were able to obtain lots and moved their houses ashore; some found refuge with relations who lived in neighboring Phippsburg and Harpswell; and some lashed their homes to rafts, drifting from place to place, in search of a community that would accept them. The state placed other exiles, including a two-year-old and his mother, into its Home for the Feeble-Minded (later renamed Pineland Center) in New Gloucester. One child, 14, was reportedly sent to an orphanage. A few months earlier, her father, a widowed Civil War veteran, had been consigned to the Old Soldiers Home in Augusta.

After the islanders' removal, Democratic Gov. Frederick Plaisted and his Executive Council, who orchestrated the diaspora, ordered any structures standing, save Malaga's red schoolhouse, razed.

Fear that the islanders might attempt to return and re-colonize seems to have triggered the state's final, ghoulish step to eradicate evidence of the Malaga settlement. The governor ordered the island graves exhumed and re-interred at Pineland. Shortly after five boxes of remains were delivered and buried, the first of at least nine islanders who were imprisoned in the institution died. Most of the others would die there, too.

Much of Malaga's history lies within the graves at Pineland. But what Randolph Stakeman, a professor of African-American history at Bowdoin College, calls "the most ruthless attack on residents" in Maine's history also speaks of an entire era—one characterized by racism, strict Victorian values and muckraker journalism. Most significantly, it speaks of eugenics — the belief, widely held at the time, that human beings could be "improved" through selective breeding.

The souls who populated Malaga for more than half a century remain the victims of myth and conjecture, gossip and hearsay. No other island community has been more written about, or more maligned; no islanders — or their descendants — have been so sorely treated and abused, so deeply stigmatized. Even today, few descendants speak of Malaga.

"It's the one Maine story told about African Americans at all, and it's not told often," explains Stakeman. "Malaga is the only instance in history where the state came in wholesale and moved people out."

"Just hearing the word 'Malaga' gives you a creepy-crawly feeling," adds Gerald Talbot, past president of the NAACP's Maine chapter and a former member of the state legislature. "It's like hearing the word 'Dachau.'"

Ignoble tales

Occupying an island that had evolved into a desirable piece of real estate for the surging summer cottage industry at the beginning of the 20th century, Malaga's small fishing community was the subject of shock-inducing "news" stories for a decade before the eviction. Published accounts of miscegenation, incest and general debauchery included descriptions of "lazy and shiftless men" and of unwashed, illiterate, mixed-race paupers living in "hovels" and "huts." Circulated throughout the state and in Boston, the reports not only bypassed fact, but ignored economic and cultural similarities between Malaga and other offshore communities. But ignoble tales, in print and on the tongues of local gossips, grew wilder over time and stigmatized the then-called "Malagoites" before and after their exile.

No one has inhabited the island since the eviction.

The word "malaga" means "cedar" in Abnaki; locals pronounce it "Malago." Today, stacked lobster traps, belonging to local fishermen, pepper the island's bold edge. Lying at the mouth of New Meadows River, about

The origins of Malaga's mixed-race settlement have been the subject of rumors and ill-researched stories for more than a century.

100 yards from Sebasco in Phippsburg, the island must look much as it did when descendants of Benjamin Darling rowed into a cove and settled on shore.

Oral tradition has it that Darling was black and his wife, Sarah Proverbs, white. But John Mosher, a graduate student at the University of Maine in Orono, whose 1991 master's thesis is the most comprehensive analysis of the colony's history to date, said he was unable to verify the marriage or Sarah Proverbs' existence. Mosher did confirm that Darling purchased nearby Horse (now named Harbor) Island from William Lithgow, Jr., for 15 pounds on July 6, 1794, and became the island's first inhabitant.

Without attribution, news articles published as recently as 1997 describe Darling as a "former slave" or an "escaped slave." In fact, nothing is known about his background. A turn-of-the-century genealogy of the Darling family offers three differing accounts of his early life, one of which centers on his mother, allegedly a slave, smuggling him out of the South. This sliver of speculation may be the source of the variously-told "concubine story," one of many myths about Malaga's early settlement which persist today.

The tale begins with the island populated by the African (or West Indian) concubines of sea captains (or Bath sailors or slave traders), who delivered the women to the island before sailing home to their families. The men, who promised to return, never came back.

Another popular narrative depicts the island as originally populated by a shipload of slaves, enroute to the South, who were abandoned on Malaga at the end of the Civil War. Yet another yarn, one of the oldest, portrays the earliest settlers as escaped slaves from the South who "lived like beasts" in underground tunnels. This tale is less often repeated today, but it may be the source of local, pre-World War II descriptions of Malaga children with small, blunted horns growing out of their heads.

The fable also may have spawned old and recent press accounts stating, again without attribution, that the island was part of the underground railroad prior to the Civil War (1861-65). According to Mosher's research, however, no one inhabited Malaga until after the war began.

Of the concubine and ex-slave stories, Mosher says, "I can't imagine why they persist. Neither the Darlings nor the Griffins, the families who first settled Malaga, have a direct connection to slavery or were involved with slavery in any way. All of them were born free, in Maine. The Darlings were one of hundreds of African-American families living in the state."

"Tales stick around because they fit people's ideas of

racial stereotypes," offers Stakeman, explaining that the "concubine story fits stereotypes of excessive sexuality" among women of color.

According to the Darling family's genealogy, housed at Phippsburg's Public Library, Ben was described as "sturdy, industrious" and he enjoyed "many staunch friends." He had at least two sons, Isaac and Benjamin, Jr., who became fishermen, married and produced nine and five children, respectively.

The Darling family owned Harbor Island until Isaac conveyed it to a Joseph Perry in 1847. Isaac migrated to Bear Island in Phippsburg and Ben, Jr., settled on Sheep Island in Harpswell.

Over the next 50 years, Ben Darling's descendants settled on Malaga and on other islands in the New Meadows River and the northern section of Casco Bay. They were joined by other fishing families of Scottish, Irish, Portuguese and Yankee descent who, like the Darlings, merged in and out of island settlements and survived as best they could, explains writer and historian William Barry.

Barry's 1980 *Down East* magazine article, "The Shameful Story of Malaga Island," was the first investigative, and humanistic, account of the community.

"Descendants of Will Black, who was called 'Trader Black,' also lived on Malaga," says Barry, explaining that Black, an African-American who settled Bailey Island, was the most famous frontiersman in the East during the 1720s.

A number of mixed-race settlements existed on the islands and the coast, notes Mosher, "but by the 1880s, they're getting ostracized on the mainland. Interracial marriage wasn't a big deal on the islands, but on the mainland, people became extremely prejudiced toward blacks and whites living together."

At the same time, tourism and the summer cottage industry had begun to fill shriveled local coffers. Wooden shipbuilding, once the bastion of Phippsburg's economy, had begun to decline following the Civil War. And over-harvesting, coupled with changing consumer tastes, had crushed the fishing industry as well.

Malaga became the "last refuge for some poor Anglo- and African-American fishing families in Casco Bay," reflects Mosher, adding that islands had been regarded as worthless until the end of the 1800s. "Some of the people who migrated to Malaga had been hopping from island to island, just one step ahead of the summer hotel and cottage industry. I can't think of any Casco Bay island that didn't have a hotel on it by 1910."

Even though Malaga residents had been self-sufficient for decades, some began requiring aid from the Town of Phippsburg in 1892, which "made social tensions even worse," observes Mosher. In addition, some of the ramshackle homes, built by a few newly arrived islanders, were beginning to offend some locals' Victorian sensibilities — and spoiling the view for newly arrived rusticators from Boston, New York and Philadelphia.

By 1900, towns along the Maine coast were fighting for possession of abandoned or un-deeded islands (Malaga appeared to be one of the latter), so they could be sold, developed and taxed. But neither Phippsburg, which had legal jurisdiction and the burden of Malaga's growing pauper accounts, nor Harpswell, which enjoyed control over islands near Malaga, wanted the community. A feud ensued between the two towns, catapulting Malaga and its residents into the headlines.

A typical article, containing quotations from an unnamed fishing boat captain, appeared in the *Bangor Daily News* on March 14, 1902: "Every winter has had the same story of starvation and suffering. ... 'They were too lazy to cut wood. ... They were wofully [sic] ignorant and chillish [sic]. Few, if any, can read or write, and many have never heard of God.'" Yet other newspaper accounts criticized the islanders, who were members of the Nazarene church in Sebasco, for being too vulnerable to religious influences.

To the horror of Phippsburg residents, who wanted no affiliation with the island and had worked hard to draw tourists to Popham Beach and environs, the legislature awarded Malaga to their town in 1903. Town fathers responded quickly, persuading lawmakers to reverse their decision and, in 1905, the islanders became, like Maine Indians, wards of the state.

"Some of the people who migrated to Malaga had been hopping from island to island, just one step ahead of the summer hotel and cottage industry. I can't think of any Casco Bay island that didn't have a hotel on it by 1910."

"The situation split the town of Phippsburg along socio-economic lines, between wealthy types and fishing families, some of whom were related to the islanders in some way," Mosher said.

Popularly called "No Man's Land" during and after the feud between the two towns, Malaga nonetheless functioned like most offshore fishing communities at the turn of the century, with some notable exceptions: Malaga lay only 100 yards from shore; it lacked a school or a formal education program until six years before its inhabitants were evicted; and it was populated by white, black and mixed-race people, a few of whom may not have been legally married. All functioned as family units, however.

In many ways, Malaga's community resembled other year-round fishing enclaves off Maine's coast. People were poor and survived mostly by fishing, lobstering and clam digging for food and income.

In addition, census records reveal, one household of three females (two sisters and a niece) was headed by a woman, Eliza Griffin, who lived in a salvaged schooner's cabin. And a number of island women, both married and single, donned trousers and fished — on their own and alongside the men — within view of the mainland and other coastal fishermen. Griffin, a laundress, housekeeper and "fisherman," was reportedly earning more money than any man on the island.

In every other way, Malaga's community resembled other year-round fishing enclaves off Maine's coast. People were poor and survived mostly by fishing, lobstering and clam digging for food and income. Some adults held seasonal mainland jobs at local resorts and farms; a few women took in laundry; some residents just didn't want to work. A number of the children were taught how to read and write at home, while others received no tutoring; some worked to help support families. The elderly appear to have cared for children while parents worked. And though Malaga boasted no formal government, the community operated under the leadership of one man.

The first known leader was James McKenney, a Phippsburg native of Scotch-Irish descent who, with his wife and a couple of children, moved to Malaga about 1870. Whether he deposed an earlier leader, no one knows. But McKenney, whom news reports acknowledged as one of the best fishermen in the area, became the island's "king" — a title derided in some quarters, then and now, and sometimes attributed to an African tradition.

Disdain of the title ignores the significance of an early island custom, well established in Penobscot Bay during the last century through Robert "King" Crie on Ragged Island (Criehaven). For decades, Crie controlled resources, such as land for development and farming, and the availability of jobs. He performed other social functions as well, such as punishing people who violated norms and mores, and ridding the community of those who didn't "fit in."

McKenney, according to newspapers (references to him were surprisingly consistent), was a well-spoken man who could read and write, built the best house on Malaga and was considered a strong, dependable leader and spokesman. Like Crie, he dispensed favors, advice and punishments; arranged for jobs; helped fishermen improve their homes and equipment; and conducted Sunday services when islanders could not get ashore to the Nazarene church. McKenney organized commerce, also, according to oral history handed down from Clinton Hamilton, a general store owner, to Ellsworth Miller of Chebeague Island, located about 10 miles from Malaga.

"They'd start out together in the morning before daylight, one to three of 'em in a dory," Hamilton told Miller. "You could hear 'em singing for miles! You could hear 'em coming before you could see 'em. About a dozen of 'em come in a group. Used to sing old Civil War songs, everybody rowing along to the music. Every one of 'em could play a fiddle. They'd get to the east end [of Chebeague] by sunrise, ready to take groceries aboard. They'd bring barrels full of bait to trade — clams shucked and salted — and then put [the groceries] in the wooden bait barrels. Everybody bartered back then. Clint said they were the most honest people he ever did business with — everybody else was trying to [cheat] him. ...

"They came whenever they needed to, maybe a couple times a year. They just got what they needed, no more. They didn't need money for anything. When they'd get groceries enough to keep them going, they'd just sit down. It must have been a nice life."

Continued on page 90

"You must be from Malago".

Seated at the dining room table outside his kitchen, Gerald McKenney listens silently as I read passages from a 1995 newspaper article: "[The Malaga islanders] built a haphazard community of shacks with dirt floors and leaking roofs. ... Couples did not marry but simply set up house, and there was much incest. Dogs and children of various colors ran freely. ... Winters, they burned driftwood for firewood, and often succumbed to sickness and death."

"That's shit," responds the 74-year-old, mildly. "Excuse me, but that's what it is. This is one of the reasons people won't talk about Malaga Island: They're afraid they'll be misrepresented by the press."

McKenney, born in Phippsburg 13 years after the state-sponsored eviction of the racially-mixed community, is the youngest grandson of James McKenney, Malaga's "king." He is also one of the few island descendants willing to speak about his connection to Malaga.

Most so-called "news" stories, like local gossip, he says, continue to exploit unsubstantiated rumors and deepen the stigma borne by islanders and their descendants almost a century after the forced exile. Stigma not only intensified social pressures on island families to melt into mainland communities, but created a legacy of silence that smothered oral history.

Throughout this century, Malaga has remained a source of derisive jokes in the Phippsburg-Harpswell area, where some islanders settled after the eviction. A mistake at the classroom blackboard or a day of poor fishing, McKenney said, can trigger gibes, such as, "You must be from Malago" (the local pronunciation). But islanders and their descendants have suffered more stinging slurs.

"At school kids teased me, saying, 'You're a half nigger' and 'Why don't you go back to Malago where your father came from.' It was confusing, because my parents were white and they were born in Phippsburg," says McKenney, whose son, 28, has endured verbal abuse also.

"When you're young, it makes you bitter, it hurts your feelings. As you get older, and if you can talk to your immediate family about it, you learn to live with it. But in some families—within my own—it's not discussed at all, even today."

McKenney's mother, Elsie Anderson McKenney, died during his birth. His father, James Irving McKenney, who moved to Malaga as a boy and, after he married, lived there until the eviction, gave Gerald to his widowed maternal grandmother to raise. McKenney says he saw his father, who later remarried, only occasionally. When he asked questions about his grandfather or the island, McKenney said his father's response was, "Don't talk about this."

"Islanders were badly used because of the stigma," continued McKenney, emphasizing it was much stronger when his older siblings grew up. Nonetheless, he points out, "a number of descendants became quite successful."

McKenney suspects the "taint of association with Malaga" may have pursued his famous grandfather in death. A talented fisherman, community leader, musician and composer, James McKenney and his wife, Salmome Griffin McKenney, are buried not in the cemetery at the Nazarene church ("where they were devoted members," he explains) but in "unmarked graves where the poor people [of Phippsburg] are buried."

A self-described "jack of all trades," McKenney grew up poor and left school in the eighth grade to help support himself and his grandmother, a midwife. "I knitted trap heads, shelled mussels, dressed fish, mowed lawns—anything I could do to make an honest dollar," said McKenney, who is retired but continues to knit as a hobby.

Of the eviction, McKenney observes, "The islanders were abused, degraded and evicted because of racial intermarriage- and poverty. They weren't animals. They were happy on Malaga—they didn't want to leave. Nobody deserved what happened there," he reflects softly. "Nobody deserved what's happened since."

Gerald McKenney

- Deborah DuBrule

THE LAND

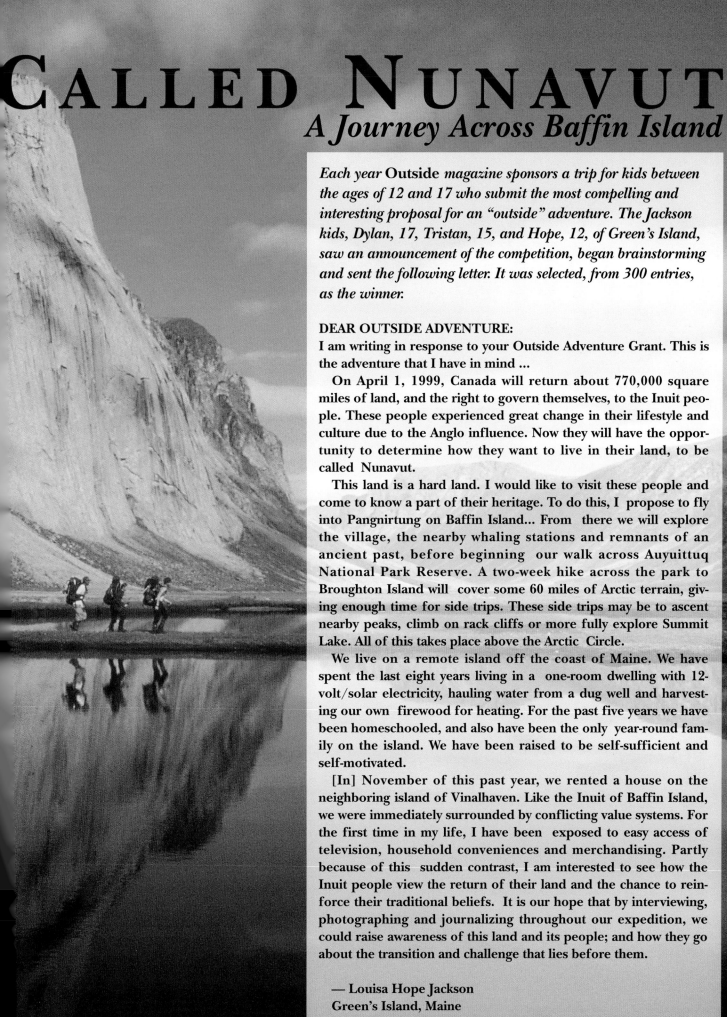

CALLED NUNAVUT
A Journey Across Baffin Island

Each year Outside *magazine sponsors a trip for kids between the ages of 12 and 17 who submit the most compelling and interesting proposal for an "outside" adventure. The Jackson kids, Dylan, 17, Tristan, 15, and Hope, 12, of Green's Island, saw an announcement of the competition, began brainstorming and sent the following letter. It was selected, from 300 entries, as the winner.*

DEAR OUTSIDE ADVENTURE:
I am writing in response to your Outside Adventure Grant. This is the adventure that I have in mind ...

On April 1, 1999, Canada will return about 770,000 square miles of land, and the right to govern themselves, to the Inuit people. These people experienced great change in their lifestyle and culture due to the Anglo influence. Now they will have the opportunity to determine how they want to live in their land, to be called Nunavut.

This land is a hard land. I would like to visit these people and come to know a part of their heritage. To do this, I propose to fly into Pangnirtung on Baffin Island... From there we will explore the village, the nearby whaling stations and remnants of an ancient past, before beginning our walk across Auyuittuq National Park Reserve. A two-week hike across the park to Broughton Island will cover some 60 miles of Arctic terrain, giving enough time for side trips. These side trips may be to ascent nearby peaks, climb on rack cliffs or more fully explore Summit Lake. All of this takes place above the Arctic Circle.

We live on a remote island off the coast of Maine. We have spent the last eight years living in a one-room dwelling with 12-volt/solar electricity, hauling water from a dug well and harvesting our own firewood for heating. For the past five years we have been homeschooled, and also have been the only year-round family on the island. We have been raised to be self-sufficient and self-motivated.

[In] November of this past year, we rented a house on the neighboring island of Vinalhaven. Like the Inuit of Baffin Island, we were immediately surrounded by conflicting value systems. For the first time in my life, I have been exposed to easy access of television, household conveniences and merchandising. Partly because of this sudden contrast, I am interested to see how the Inuit people view the return of their land and the chance to reinforce their traditional beliefs. It is our hope that by interviewing, photographing and journalizing throughout our expedition, we could raise awareness of this land and its people; and how they go about the transition and challenge that lies before them.

— Louisa Hope Jackson
Green's Island, Maine

P.S.:

Hope:
Forty-five percent of the Inuit population is under the age of 15. As young people ourselves, my brothers and I have a distinct advantage in communicating with other young people. We do not expect it to be difficult to find kids, nor do we expect any problem in making friends among them. I believe we share a common factor in that we, too, have grown up in a beautiful, harsh and isolated place. Growing up the way we have, we are perhaps more skilled than some children at making immediate, though sometimes brief, friendships.

Anyone can talk to government officials and tribal elders, but to our knowledge no one is talking to the kids. We want to know what it is like for them to live there.

Dylan:
This expedition we have planned is unique in that it is in an area very few people have ever heard of, let alone seen. The land is largely unaltered by human presence; last year only 1,000 people visited the park. I have always enjoyed the outdoors, in particular the wilderness, and I am attracted to the idea of going to such an isolated, unspoiled place.

Tristan:
In summary, we plan to study by being keenly observant and through meticulous record keeping. The goal is the education of our group and the sharing of this information on our return.

"We do not expect it to be difficult to find kids, nor do we expect any problem in making friends among them."

Preparations

Hope:
April 2 to 21, 1998
The last two weeks have been very busy. On the morning of April 2, John Alderman from *Outside* called to ask questions about us and the adventure we had proposed. That night he called again, this time to tell us that we had won the 1998 Adventure Grant! When no one said anything, John asked "Well, aren't you excited?" We answered with unanimous gibberish; we were so excited we couldn't speak!

When we get our packs and shoes, we will begin hiking around the island with bags of sand to get in shape for the hike. Right now we are still sort of pulling things together, and it's all a little hectic, but starting to settle down.

We have been gathering information for the last two weeks. The process up to this point has been very educational; we've had to do a lot of writing, making telephone calls and researching. We have been enjoying this part of the adventure, but we are definitely looking forward to the actual hike. It's going to be amazing!!

Dylan:
I was nervous about a TV station coming out to our home on Green's Island. One of the things I love about the island is the wonderful privacy we have there, not that we're hermits or anything, but I wasn't sure how I felt about our home being broadcast to midcoast Maine on the six o'clock news. In the morning, before the camera crew came out, Tristan, Hope and I had a quick talk to try and figure out what it was we were going to say to them. The meeting dissolved into us warning each other not to say anything too stupid. I think we did okay...

Karen Jackson:
This being Mother's Day, I thought perhaps I could get away with writing the kids' weekly journal entry ... it has been a quieter week than the few previous, and I must admit it's been nice. For the first few weeks after Dylan, Tristan and Hope were chosen for their expedition, we jumped with both feet into a world of fax machines and the internet, TV and newspaper interviews. Quite a sudden change from an uneventful winter, since our life

"All the kids in Pagnirtung are happy because this is the first cargo in three months and now the northern store will have Pepsi again."

on the islands has been pretty much out of the public eye, until now ... it certainly added some spice to our usual routine.

Hope:
I had talked to Del so many times on the phone that when she came it felt like I already knew her! I think that we are all going to get along really well on this trip, and probably by the end of the trip we will feel like we have always known each other.

Dylan:
June 13, 1998
Yesterday our backpacks came! These things are pretty neat; high-tech super packs with all sorts of cool little features. Comfortable and tough. It's going to be wonderful to hike with these packs on. We also received very nice daypacks, down sleeping bags and three tents. All of it is spread around the office so that we can paw through it and look at our equipment.

June 15, 1998
Today more equipment came: shoes and sandals from Timberland. Man! We're going to be hiking in style! Two months ago, when we put together our gear list, I never thought to imagine what it would all look like on my back, on my feet and piled up in the office. It really does look like we are preparing ourselves for a polar exploration! I can picture one of the great arctic explorers sitting in his living room surrounded by a mound of gear, saying to himself, "Did I really order all of this stuff?" I must admit thought, having all this equipment makes you feel as if you could take on anything!

Traveling North

Dylan:
July 22, 1998
Up at 5:30 a.m. today... The best aspect of the flight was the captain's translations. While describing the safety features the fact that English was his second language became apparent. After telling us what to do in French, and Inuktituk, he let the Americans know that we should tighten the seatbelts around our tiny lips — oops, that was supposed to be hips. If we had small children with us, he felt sorry for us. But we should put on our oxygen mask first, then help them; the same if we had those who acted like small children with us!

Tristan:
July 25, 1998
Today is the Nunavut day celebration ... I think there was a storm on the real Nunavut day (July 7). Anyway, games are going on all day, plus more feasting. Today's new dish: I tried the fat and skin of a walrus killed three months ago and allowed to age for full flavor. Personally, I'll stick to the fresh maktak... [the skin around the blubber of a bowhead whale].

"We found out yesterday that a bowhead whale was killed ... today they are going to bring all good eating parts to the landing at about 7 p.m. ... Practically the whole town was there to see the hunters come home."

Dylan:
We found out yesterday that a bowhead whale was killed on an island near Pangnirtung. And today they were going to bring all good eating parts to the landing at about 7:00 p.m. We were in our tents resting and heard lots of people cheering and talking. So we went over to the landing and practically the whole town was there to see the hunters come home...

Hope:
July 25, 1998
There were tons of people at the feast, and tons of whale pieces. There were a lot of speeches, but I couldn't understand anything they said. We all ate maktak and fried dough and cookies. There was more talking, and then we all went up to the Hamlet Center to dance. For music, there was a drum, the spoons, the accordion and the bagpipes. I didn't dance because I was too tired, because by that time it was 10:30 p.m., but Dylan and Tristan danced almost the whole time. The dancing went on until 1:30 a.m....

Dylan:
Reflections: Hunting is a way of life for the Inuit. It has provided food, clothing and, at times, a large part of their income. Today it provides a tie to the past and a reason to spend time on the land. After seeing the prices of food up here — $9 for a pound of powdered milk — I understand why the Inuit choose to take advantage of this free, abundant and easily accessible source of food. From what we've seen, people here do not hunt for sport. Every part of every animal is used. The scraps from a butchered ringed seal wouldn't fill a one-gallon bucket...

As tourism becomes a larger part of Nunavut's economy it can be expected that outsiders' opinions will have more of an influence on how the Inuit live their lives. A hunter that I spoke with said that visitors are often disgusted at the sight of a seal being butchered on the beach. "I tell them to think of the last time they ate a hamburger. I tell them it's not so different, at least this seal was free before he died." I see the hunter's point, but many other people don't.

Hope:
Reflections: Four days ago, Pangnirtung was celebrating its first bowhead whale kill since 1946. There were over 230 strips of baleen in its jaw. The elders of this hamlet were given a pair each. A few days ago, we feasted alongside the rest of Pangnirtung's residents on a 40-foot long table covered with whale blubber. Older Inuit women dressed in traditional clothing cut the skin off the blubber. The black skin, high in vitamin C, is the prime cut.

Today, there is a more modern celebration in progress. The cargo ship ALVIT is in the bay. All day, D-9 tractors unload freight. Julia had told me that the kids in Pangnirtung are happy because this is the first cargo in three months and now the northern store will have Pepsi again...

I wonder how many people here have noticed that fiberglass insulation and bowhead whale blubber are the same color? Both serve to keep Inuit families warm... And today for me they represent all that is the new Nunavut territory.

Auyuittuq National Park Pangnirtung Fjord, Overlord Camp

Tristan:
July 28, 1998
Last night Jamie [our 17-year-old Inuit companion] decided that he didn't want to attempt the hike. I don't think that this decision was an easy one for him. From the time he arrived we could tell that, as much as he wanted to hike through the park, he also had a strong desire to spend the summer with his family on Broughton. This indecision worried the rest of the team because if there is one thing that an expedition requires, it is total commitment. I think that Jamie would have made an excellent member of the team and I am sorry to see him go.

I will never forget this, our first day camped in the park. When we first landed at Overlord at the very end of the Pangnirtung fjord, clouds were settled low, hiding all landmarks. As the morning progressed, the clouds started to break up and we caught occasional glimpses of the 3,000-foot peaks that towered over us. To the east sits Mt. Overlord and to the north Twenweather glacier sends stubby fingers of ice between the ridges. West of us lies another line of ice and snow-capped mountains.

Tristan:
July 29, 1998
Day 2 of trek — another tragedy struck today. All started well; we had breakfast at seven and were all ready to load up the packs and start hauling our loads north when Julia surprised us all by announcing that she didn't want to go on. We all tried to convince her to stay, but her mind was made up. Yesterday's walk, she said, was too much. And while we volunteered to carry part of her loads, we knew that there were much harder hikes to come. The tough part now was evacuating her from the park. We called a boat on the emergency radio and Del ended up going into Pangnirtung with Julia. Our plan is to chill for the day and be ready to go whenever Del can make it back from Pangnirtung — hopefully tonight or tomorrow morning. Toward evening, a 30-knot wind kicked up and there was no sign of Del. Somewhat discouraged by the loss of another member of the team and the ensuing delay, we all retreated back to our tents early. Only Tommy, who we didn't know anything about until we got here, is left to represent the Inuit.

Hope:
Today at breakfast in our Overlord camp, Julia began to cry. She is so quiet that it was difficult to hear what she was saying. But her tears spoke clearly. She wanted to return to her home on Broughton Island. What a huge disappointment. Especially after Jamie left the expedition yesterday because he didn't think that he would be able to make it. Jamie was afraid he'd quit once he got in the field.

Reflections: There will be plenty of time for the team to sort out the lessons of our attempts with these two Inuit teens. Right now, they are focusing on the expedition and our remaining Inuit companion, Tommy. But questions about Jamie and Julia are running through my mind. Did we make them feel welcome? Were we clear in describing the commitment it takes to hike across Auyuittuq Park? Could the Jacksons have spent more time in the initial interviews? Regardless, we are now a group of six.

July 30, 1998
Auyuittuq National Park: It is very windy today. The worst is when the trail is sandy. Today we got sandblasted the whole way. It really hurts to be hit in the face by sand flying at 40 miles per hour. We crossed the Arctic Circle today and also got to see

"We are surrounded by cliffs and mountains... I've never seen a place like this before."

Shwartzenbach Falls, an 1,800-foot waterfall that lies right on the Arctic circle. There is lots of grass and fireweed here and on the way to our cache at Windy Lake. We are surrounded by cliffs and mountains... I've never seen a place like this before.

August 6, 1998
We have made it to our halfway point, Summit Lake camp site. It was wet, foggy and dreary all morning. We took a rest day and I spent lots of time sleeping, reading and resting. I have lost total track of time up here...It's hard enough remembering what month and day it is, let alone the time.

August 9, 1998
Today we hiked up and down, up and down, over big old glacial moraines for five and a half miles. A few times we would stop and say "Tommy, where in the world are we?!?" Tommy would whip out his map and say, "I think I've got it. Yep, we're on Baffin Island!" But Dylan, Tristan and I disagreed. We thought it looked a lot more like Mars than Earth. And it did...

Dylan:
August 18, 1998
We have been in the park exactly three weeks and we are all eager to get the boat to Broughton. The boat was hours late, but when it arrived, it was obvious why. There was a seal draped over the stern and another in plastic bags on the deck. The ride was only an hour and a half long.

---

*Hope, Tristan and **Dylan Jackson** live on Green's Island near Vinalhaven.*

THE DEFENSE THAT FAILED

Had Ann Brown's murderer been tried 30 years later, the verdict might have been different

RANDY PURINTON

Illustration by Siri Beckman

Ann Brown never finished cleaning the breakfast dishes. She was easily overpowered and died quickly. It took but one cut to sever her carotid artery and windpipe along the right side of her throat. She stopped breathing instantly and blood immediately stopped flowing from her heart to her brain. Later, a witness recalled that her blood flowed six to eight feet from the fatal wound. Much of the news in Belfast's *Republican Journal* on that day was about nominations of candidates for the Presidency, but on Islesboro, April 16, 1856, the talk was about the murder of Ann Brown by her husband, Joseph.

A broad view of the events related to the murder are recorded in John Farrow's *History of Islesboro,* published in 1893. But a more detailed account about the Brown family history and Joseph's trial is recorded in April 1856 issues of the *Republican Journal.* Combined, these two sources recall the story of a grisly family tragedy with overtones of gothic horror.

Ann, whose full name was Queen Ann Brown, was a daughter of Simon Dodge, who moved to Islesboro from Block Island, Rhode Island. He married Polly, the daughter of Sylvester Cottrell, one of two early families who settled on Hewes Point. Ann was born on Islesboro in 1825, when Simon was 57. Simon, wrote Farrow, "died in his field" a year later.

What we know of Ann's character can be gleaned from trial testimony. Joseph Brown's sister Pamelia described Ann as a "spleeny woman" who was always complaining about her health and had no set time for going to bed. A person with "spleen" was regarded as irritable and spiteful. Pamelia had lived with Joseph and Ann during the previous February and March, possibly to help care for the family during the illness of an infant daughter, Martha, then about four months old. Her testimony could describe any relative who has lived with in-laws during winter months.

Joseph's daughter Pamelia testified that Ann was always kind to her father. Farrow, in his *History,* wrote that Ann was "an entirely inoffensive woman." A reporter for the *Journal* wrote that Ann was "small, rather feeble, an inoffensive woman, prepossessing in her looks." This describes a polite, slightly built and frail woman with an appealing face. As a mother and householder, Ann suffered through the loss of four children. She taught her daughter Pamelia to care for the infant, Martha, and Ann seems to have kept a clean house. Though Joseph for years accused Ann of infidelity, it's not likely that Ann could have had an extra-marital affair without a neighbor, relative or even her daughter noticing. She was visited by friends and visited others regularly. Those with whom Joseph shared his suspicions refused to believe him.

Ann did have a chronic medical problem: syphilis. This information was provided by a family doctor, John Paine. In response to questioning by the prosecutor, Paine said he had diagnosed Ann's illness a year or so before the murder. On the witness stand he did not at first want to disclose the name of Ann's disease, calling it "one of the unmentionable diseases." But after further questioning he went on to admit that Ann's case was one of the "worst forms" of the disease.

He said Joseph challenged the doctor's diagnosis but he paid for a prescription anyway. When asked by the prosecutor if Brown was concerned about the origin of Ann's infection, Paine said that Brown revealed nothing. An infection may be the reason why Ann often complained about her health and may also be the reason why the infant Martha was so ill that Joseph's sister moved in with the family. Assuming that Paine's diagnosis was correct, it is not likely that Ann was infected by a secret lover. Joseph's reluctance to talk with Paine about the origin of Ann's infection, perhaps, said enough.

Joseph was the fourth of nine children born of the marriage of John Brown and Peggy Hewes. Peggy died in 1840 but John lived to learn that his son was a murderer. By his 38th year, Joseph Brown had established himself as a family man and property owner. He had always worked hard for what he owned, and though some suspected that he drank too much at times, it does not seem to have prevented him from fulfilling family and business responsibilities. Joseph was the master of a sailing vessel that often shipped freight from Belfast. He also sailed up the Penobscot River to Bangor. Some spoke of an occasional wildness in Joseph's eyes and a concern that Joseph was too often "broken of his rest." His relationship with his wife, Ann, was outwardly stable and cordial, but it was known that he struck Ann at times. The night before the murder, Ann would not return to the house with Joseph because she was afraid of him, and she insisted that someone accompany her and spend the night.

By late March 1856, it seems as if fatigue, years of jealousy, lack of sleep and maybe alcoholism and syphilis had taken their toll on Joseph. Three weeks before the murder, he entered his brother William's house and began accusing his wife, Mercy Jane, of being a harlot. William told him to leave but Joseph seized him by the collar and laughed. Abruptly, Joseph left the house and William's son locked the door. Seconds later, Joseph smashed in the door and seized William again, saying he was going to beat him for nonpayment of a debt. William grabbed some tongs and prepared to defend himself. He noticed that Joseph's eyes were rolling and that froth was collecting at the corners of his mouth.

Then Joseph calmed down, apparently because a neighbor, Thomas Fletcher, intervened. Joseph asked William for some cistern water but William refused to allow him to pass through the room where Mercy Jane had sought refuge. William offered his brother well water but Joseph refused, insisting that only cistern water would do. After this exchange, Joseph left the house and began sprinting across William's field. He stopped suddenly and began walking in a large circle. After having created the imaginary circle, Joseph walked inside it and held his arms over his head. Witnesses claimed Brown was acting "after the fashion of incantations, to get rid of evil spirits," and testified at the trial that they thought he was crazy.

Though Joseph's behavior at his brother's house could be attributed to mental illness aggravated by alcoholism, syphilis may have contributed to his behavior as well. Dr. Paine testified that Ann had a "worst form" of the disease, but syphilis can affect individuals differently. Indeed, Joseph may have been suffering from a "worst form" of the disease, too, though his symptoms could have differed from Ann's.

In the earliest stage of the disease, a person usually develops ulcers on the skin. In the second stage, lesions form on the mucous membranes accompanied by headaches, sore throat and rashes. Lesions in his mouth and a sore throat might explain why Joseph asked specifically for cistern water at his brother's house; he had learned from experience that cistern water might be more soothing than straight well water. Tertiary syphilis is the last stage of the disease. Symptoms in this stage include loss of balance, dementia or insanity, blindness and congestive heart failure. About one in four people with untreated syphilis will advance to the tertiary stage. During Joseph's trial Dr. Paine recounted that a week before the murder he saw Joseph stagger to such an extent that he hesitated to accompany Joseph to his house. This stagger might have been caused by partial paralysis. The doctor did not suggest that it was caused by drunkenness.

There was no cure for syphilis in the mid-1800s. The prescription that Joseph bought from Dr. Paine was probably mercuric chloride, a drug that was believed to be effective in bringing a cure. Martha, Ann's infant daughter, could have been infected while in the womb — bacteria can pass from mother to child through the placenta. Pamelia, on the other hand, had avoided illness, perhaps because she was born before Ann and Joseph were infected.

So on Islesboro in the spring of 1856, there was a family of four in which three of its members were suffering through various stages of syphilis. The father became infected during a tryst in a coastal city. His case was the worst, because it attacked his brain. Alcoholism only made matters worse. The mother contracted the disease from her unfaithful husband. Her case was less severe but was serious nevertheless. The infant daughter was infected before she was born. Her fate is unknown. The uninfected daughter grew to adulthood and married, but she lived with the memory of her father slashing her mother's throat.

"Farewell, house and wife"

Thomas Fletcher's testimony relating to the order of events on the day before the murder can be a challenge to follow. It seems that Joseph and Ann were visiting with Thomas and Penelope Fletcher at Hewes Point. During this gathering, Joseph's suspicions about Ann's infidelity again surfaced; this time he accused her in the company of the Fletchers. Brown had spent Monday night in Belfast because he had taken on some livestock meal to be delivered to Islesboro. On that night Ann and her daughter Pamelia slept in the same bed at home together. Pamelia testified that she and her mother were alone. But Joseph, who wasn't even home Monday night, asked her who the man was who she slept with that night. Ann responded that "if he saw anybody it must have been the devil."

Fletcher testified that this was the second time on Tuesday that Brown had asked Ann about a man with whom she slept on Monday night. That morning, soon after Brown's return from Belfast, Joseph asked Ann "what man it was in bed with her that threatened to bite her ear off." At the end of their visit with the Fletchers, Joseph asked Ann if she were returning home with him but Ann refused to go unless someone would accompany her and spend the night. She said she was afraid of Joseph. Penelope volunteered and spent the night at the Browns' house. Thomas Fletcher recalled in his testimony that on that same day Penelope, concerned about Joseph's physical and mental condition, questioned Joseph about his health and urged him to stay for dinner. Joseph pushed her away but then he reluctantly sat at the table. He ate very little and was soon on his way. Sometime before Brown returned to Belfast he wrote a message on a door in his house, "Farewell, house and wife, and blessed little children; may God bless you. Joseph J. Brown."

The message read more like a suicide note than a threat of murder. Regardless, it contained a tone of impending doom. Joseph seemed to have been preparing himself to meet death, his own or someone else's.

Joseph returned to Belfast and spent Tuesday night there. He might have wanted to load the remainder of the livestock meal that he brought to Islesboro on Monday. The *Republican Journal* does not record what he did Tuesday night, but Farrow's *History of Islesboro* states that, "The day before the murder Brown had been to Belfast, and, as usual when there, he had indulged freely in intoxicating liquor."

None of the witnesses at Brown's trial, however, ever suggested that alcohol caused problems in Brown's family or his business. A defense witness suggests "an addiction to excessive stimulants," but that could have been a reference to tobacco. Not even Brown's daughter, who testified against him, speculated that alcohol was a factor in the murder. Brown's lawyers argued that there was a history of insanity in the family. This, combined with deep resentment against his wife, lack of sleep, a bout of heavy drinking and symptoms associated with syphilis, may all have contributed to Brown's condition on Wednesday morning. Whatever the causes, when Joseph entered his house that morning after returning from Belfast, he was inclined to commit murder if the opportunity arose.

On the morning of the murder, Penelope Fletcher left the Browns' house around six or seven o'clock. She was concerned because she had seen the message that Joseph had written on the door and she wanted to tell her husband, Thomas.

About an hour later, Joseph returned from Belfast. During her testimony Pamelia Brown, Joseph's 12-year-old daughter, recalled what happened next: she had been rocking her infant sister in her cradle and Ann was washing the breakfast dishes on a table when Joseph returned home. He came through the door, took out his pocketbook and handed it to Pamelia. He approached Ann, who told him to stay back because she was afraid of him. Joseph picked up a large knife that lay on the floor, presumably by the stove or fireplace, a long sharp butcher's knife that he sometimes used to make shavings to kindle the fire. He grabbed Ann, who fell back onto a chair next to the bureau onto which she had been stacking clean dishes. Ann begged Joseph not to kill her. At this point Pamelia tried to restrain her father, but her fingers were cut badly in the attempt. The bureau tipped, dishes, bottles and a large bowl of sugar smashed onto the floor and Ann fell while Joseph clutched her. As Ann lay on her back on the floor Joseph slashed her throat once and it was over. Ann's head lay on a pile of broken dishes. Joseph left the house.

It might have been that, after the murder, Joseph had in mind to throw himself off a nearby cliff in an effort to commit suicide. The *Journal* records testimony from a witness who said that Brown was about to jump, but decided not to when he saw his father in the distance and feared that his father might follow him over the cliff. Pamelia testified that her father returned to the house after about 15 minutes. Certain that his wife was dead, he retrieved the knife and left the house again, walking rapidly towards the shore.

By this time Thomas Fletcher was walking toward Brown's house. Penelope had told him about the message on the door, and he decided to check on the family as well as deliver some milk for the baby and meal for Joseph's hog. He met Joseph on the way and said, "Good morning." Without stopping, Joseph told Thomas to "go up to that house for there is trouble there." They parted ways — in a minute or two, Thomas Fletcher was looking at Ann's body in the corner of the house and reading the message on the door.

As the word of the murder spread, Joseph Brown was a fugitive rowing across east Penobscot Bay. A boat carrying several islanders was in pursuit. Joseph pulled in to Beech Island and found a heavy rock that he was going to tie to a rope around his neck and use as a weight in an effort to drown himself. Then he reversed his course and decided instead to face his pursuers. They met near Barred Island. Here, Joseph made a poor attempt to drown himself and was hauled into the chase boat. During the return to Islesboro, Joseph confessed his deed to his captors saying, "Yes, I killed her but she killed me first."

"A mental delusion or hallucination"

Joseph was indicted for first-degree murder. He admitted having murdered Ann, but he pleaded "not guilty" on grounds that he was momentarily insane and therefore did not plan maliciously to kill his wife. His act, his lawyers argued, was the consequence of an "insane impulse" provoked by a "hereditary predisposition to insanity." Had he been rational that morning he would not have killed Ann.

The defense called a Dr. Harlow as an expert witness. He testified that, "Jealousy is very often a symptom of insanity." By dwelling upon a particular subject a person might, through the influence of hallucination, do an insane act — "a mental delusion or hallucination leads a person to believe what does not exist." Dr. Harlow was suggesting that Joseph's long-standing suspicion of Ann's infidelity could have created a hallucination, such as the man who said he was going to bite Ann's ear off or the man in Ann's bed the Monday night before the murder, when John was absent.

"A person may be insane upon certain subjects and reason correctly upon others," Dr. Harlow went on to say, trying to explain to the jury how Brown could be a responsible person and still be provoked to violence against others and against himself. Dr. Harlow added that a delusion combined with a "hereditary tendency" toward insanity could create "an ungovernable impulse to commit homicide upon the instant."

Other witnesses pointed to evidence of insanity in Joseph's family, particularly to when his grandfather Paoli Hewes attempted suicide by cutting his own throat. Though John Brown, Joseph's father, never exhibited symptoms of insanity, Dr. Harlow claimed that, "The hereditary disease sometimes passes one generation and develops itself in the succeeding one ... Everybody whose grandparent committed suicide has a germ of insanity in him, and is liable to commit suicide or homicide whenever circumstances occur to call it." Joseph's brother Oliver testified that he himself had experienced insanity in the form of prolonged periods of blackouts.

It may seem as if the prosecutor handed the defense a gift by revealing Dr. Paine's diagnosis of Ann's illness. With that information the defense could have argued that Joseph was insane because he was suffering from neurological damage caused by syphilis. Ann's illness might also suggest that she, not Joseph, could have been the origin of the infection, giving at least some credence to Joseph's suspicions of infidelity.

At this time in New England cultural history, syphilis was held in such contempt that to just talk about it was taboo; Dr. Paine hesitated to even say the word. To avoid bringing more contempt upon their client, Joseph's lawyers resorted to the temporary insanity defense; insanity caused not by syphilis but by an inherited condition, thereby presenting Joseph as a victim of circumstances rather than as a calculating, vengeful predator. It took the jury only one hour to convict Joseph Brown of first-degree murder.

The *Journal* reported that Pamelia Brown, Joseph's daughter, provided the most memorable testimony of the trial. Farrow wrote in his *History* that Brown regretted even attempting a defense and was thankful that the harm he caused was not worse. From these accounts it seems as if Brown was resigned to accepting that justice had been done.

Still, he resented those who testified against him. When Pamelia came to visit her father in his cell after the conviction, he was cool to her and said he ... "didn't want to talk to anybody who had sworn his life away."

"The writer joins in that opinion"

Most if not all of the people familiar with the tragedy were satisfied that justice was served by the conviction and sentence. An editorial in the *Journal* commented that "it is not out of place to observe that this verdict was almost universally expected, and we have not heard the first person suggest that it was wrong." The people of the Penobscot Bay area were confident that their legal system worked.

Ann was buried in a small family cemetery on the east side of Islesboro. A small marble block and primitive slate headstone with the single name "Ann" carved into it mark her grave to this day. The cemetery has been neglected, and the forest is thick around it now.

Farrow wrote that Joseph was sent to the state prison in Thomaston to do hard labor and endure solitary confinement until the day of his execution. However, the ledger of incarcerations at Thomaston reveals no record that Joseph was ever an inmate there. Nor was Joseph an inmate at the asylum in Augusta. The *History* says that in September, 1856, four months after the murder, Joseph took his own life by cutting his throat with a piece of glass, but where this happened is uncertain. Brown's death certificate is not filed on Islesboro.

At the end of his account of the murder, Farrow wrote that, "The prevailing opinion at the present time [1883] is that [Joseph Brown] was insane at the time the murder was committed, and the writer joins in that opinion." With the passage of almost three decades, people had become less convinced that Joseph was able to restrain himself that awful morning. Had Joseph had the impossible opportunity to wait almost 30 years for his trial, he might have been found "not guilty" by reason of insanity.

***Randy Purinton** is chaplain at the Williston Northampton School, an Islesboro summer resident and a longtime student of island history.*

A FITTING DESTRUCTION

THE WHALING STATION AT HAWKE HARBOUR, LABRADOR

JOHN BOCKSTOCE

Photography by Nicholas Whitman

THE SILENCE OF HAWKE Harbour, on Hawke Island, Labrador, is powerful, almost overwhelming. In the summer stillness a soft fog washes over the spruce-topped granite hills surrounding the anchorage. Outside the harbor's heads, past the rocky ledges, seabirds rest on the slick tops of icebergs and swoop and dive above pods of minke and humpback whales. Ashore, the silence is broken only by the swash of waves amid the seaweed-covered rocks and the occasional eerie call of a loon.

Amid this serenity is a jarring reminder of a half-century's noise and violence. The head of the harbor is a mass of wreckage. Rusty tanks, smokestacks, conveyor belts, boilers, firebrick and twisted metal rails lie in a wild confusion. At the water's edge decomposing whalebones - vertebrae and ribs - are mixed with exploded bits of harpoon heads and small railroad wheels. Above, on a low granite bluff, are the tortured, imploded remains of two large oil storage tanks. Below them, great black streaks of charred oil smear the cliff face.

A couple of hundred yards away the half-submerged hulks of two catcher ships list against one another at unnatural angles. On both sides of the hulks five pairs of large holes pierce bulwarks - and testify to the violence of the whaling industry. These are fairleads, through which lines were run from the whales' flukes. After the whales had been killed with exploding grenades fired from the ships' harpoon cannons, the carcasses were collected and towed to the station. There the whales were winched up a slipway to the flensing deck, where workers with long-handled knives stripped off the blubber and meat and separated the bones. The pieces were then drawn up a conveyor belt for processing: the blubber was rendered into oil in "steam digesters" and the bones were ground up for fertilizer.

Commercial whaling began in North America early in the 16th century, not long after the first explorers and fishermen had reported great stocks of fish and whales in the waters off Newfoundland and Labrador. Basque whalemen set up stations on the Labrador coast near the Strait of Belle Isle and immediately produced a valuable harvest of bowhead and right whale oil for the European market. They were so successful that in the latter part of the 16th century, a fleet of 20 supply ships served half a dozen stations in Labrador.

The intensive hunting may have severely reduced the whale populations in the strait. In any case, at the very end of that century the discovery of even richer whaling grounds near Spitsbergen, north of Norway, drew the commercial whalers there. With the exception of a small American pelagic whale fishery in the 18th and 19th centuries, the Labrador coast was essentially left to fishermen, seal hunters and fur trappers.

Events in Norway were, however, to end this. In 1868 the modern whaling industry was born when Svend Foyn and his Norwegian associates began using steam-powered catcher ships fitted with heavy guns firing grenade-tipped harpoons. These innovations allowed them to take the larger, faster-swimming whales that hitherto had been very difficult to capture.

Whale processing operations at Hawke Harbour began in 1905 and ended in 1959, when a spectacular fire destroyed the station. Today, rusting ruins are all that remain.

After quickly depleting the stocks in northern Norway, the Norwegians cast their eyes further afield and perceived new opportunities in the western Atlantic. With Newfoundland investors, they incorporated a shore whaling venture and began operations in 1898, catching 91 whales that year. By 1904 there were 10 catcher ships serving 14 stations. That year the combined effort resulted in the capture of 1,275 whales.

The Hawke Harbor station began operations in 1905 and operated on and off, depending on market conditions, until 1959. The station took a wide variety of whale species: in 1928, for instance, it processed 59 blue whales, 280 fin whales, 15 humpback whales, 28 sei whales and 22 sperm whales, for a total of 399.

Hawke Harbor's final, appropriately apocalyptic moment came on Sept. 12, 1959, when a spectacular fire swept through the station, destroying everything. Today the profound silence of Hawke Harbor suggests a great battlefield - Gettysburg, Culloden Moor, Omaha Beach. There are ghosts here, and the presence of mortal contest and slaughter is palpable.

The Canadian government banned commercial whaling in 1972. Since then the only Canadian harvest of large whales has been a few bowheads, taken by the Inuits for subsistence.

John Bockstoce *is author of several books and monographs on the whaling industry in the western arctic.*

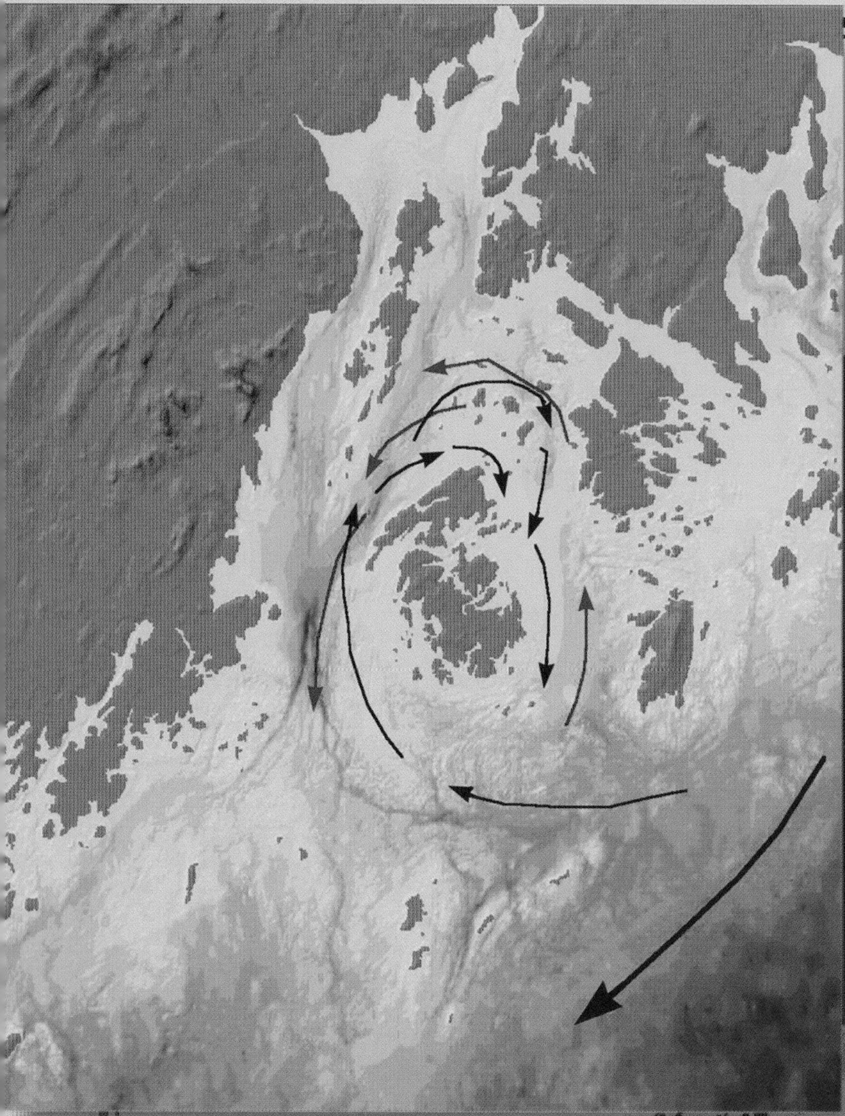

SEA SURFACE TEMPERATURE
This image was derived from an August, 1995, Landsat scene. It represents relative sea surface temperature, which can correspond to currents and oceanic fronts.

cooler   warmer

CIRCULATION
The purple arrow represents the Eastern Maine Coastal Current. The blue arrows represent circulation near the surface and the red arrows indicate patterns nearer the seafloor.
(Data derived from the work of Neal Pettigrew, University of Maine School of Marine Science.)

Layer by layer, Penobscot Bay reveals itself

PHILIP W. CONKLING

THE GREAT LOBSTER COLLABORATION

Images: C. Wentzell-Brehme, Island Institute

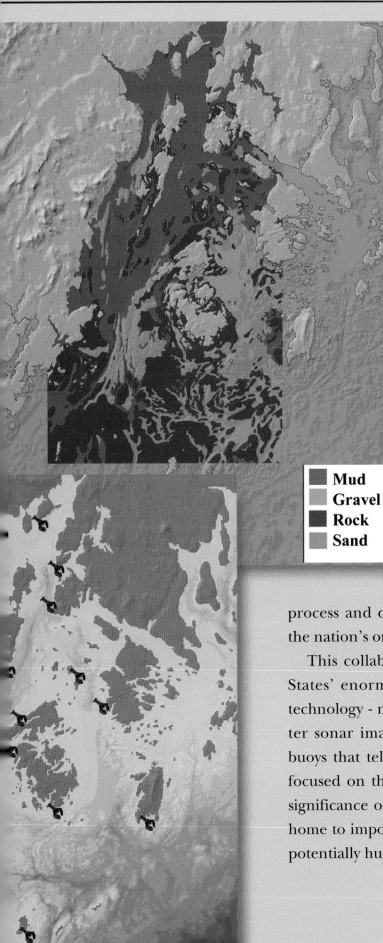

SEAFLOOR GEOLOGY
This map highlights the work of geologists Joseph Kelley and Stephen Dickson, who used bottom sampling and sidescan sonar to map the geology along the floor of Penobscot Bay.

- Mud
- Gravel
- Rock
- Sand

INTERTIDAL AND BENTHIC LOBSTER SAMPLING LOCATIONS

FOR THE PAST three years, Penobscot Bay, the largest embayment on the country's Atlantic Coast north of the Chesapeake, has been the focus of a great scientific collaboration. Its purpose is to decipher how the bay's complex ecosystem affects the patterns of abundance and scarcity for the region's lobster fishery. The project, which involves more than 100 fishermen, scientists and managers, is sponsored by a federal agency, the National Oceanic and Atmospheric Administration's National Environmental Satellite Data and Information Service (NOAA-NESDIS). This agency's mission is to collect, process and distribute vast quantities of information from the nation's orbiting satellites.

This collaboration marks the first time that the United States' enormous national investment in remote sensing technology - not just sensors on satellites, but also underwater sonar imaging capabilities and, more recently, ocean buoys that telemeter real-time data back ashore - has been focused on the problems of managing inshore waters. The significance of such work to a place like the Gulf of Maine, home to important fisheries and prone to human impacts, is potentially huge.

The Penobscot Bay project's origins lie in the Magnuson-Stevens Reauthorization Act, federal legislation that, among other things, called for deep reforms in the way this country manages its fisheries. Here in the Gulf of Maine and on neighboring Georges Bank, it has been painfully obvious for some time that efforts to manage fisheries have failed. The Magnuson-Stevens amendments (also known as the Sustainable Fisheries Act) require the National Marine Fisheries Service to apply "ecosystem" management principles to the nation's fisheries.

Oceanic dynamics are difficult to see at the ocean's surface. They change from year to year, season to season and sometimes day to day after storms or floods. The ocean is different from a forest, where you can actually count the trees to learn how many are cut or claimed each year by fire, wind, insects or disease.

We have tried for half a century, but we haven't been able to keep track of fish populations in the ocean. Today that is changing: near-instantaneous, high quality, remotely-sensed information has the potential to revolutionize the ways to approach fisheries management. Such information enables us, among other things, to focus on the details - where and when fish gather or the characteristics of spawning and juvenile populations.

DIFFERENT QUESTIONS

So what could be done differently to manage Penobscot Bay based on its ecological dynamics? With respect to the lobster fishery, for example, wouldn't the location or habitat of the broodstock - those large female lobsters that carry most of the eggs - be important? Similarly, where do these females' tiny young, the larvae that spend their first month and a half of life floating in the water, get carried on coastal currents?

New assumptions need to be made. For example, broodstock females might not necessarily be distributed uniformly across their range, which from the point of view of federal fisheries managers extends from Long Island Sound to Eastport, Maine. Also, perhaps not all eggs are created equal. Larger, older females carry a much higher number of eggs that are more likely to survive than eggs from young females. So, it is important to find out if large egg-bearing females live in disproportionately higher numbers in certain parts of the lobsters' range.

New priorities need to be set. Breeding mothers must be protected. Fisheries managers and scientists need to know how and where the young of the year are "recruited" - in other words, find the location of their

AVHRR satellite data processed by Andrew Thomas, University of Maine School of Marine Sciences.

Bob Coombs

nursery and juvenile grounds. Questions about how these little lobsters become residents of an area or bay, and whether the pattern varies when conditions are different, need to be answered.

Money should be spent to study and understand the lobster's life cycle. And because lobstermen have been an inseparable part of the system for most of the past two centuries, the differences in the way these fishermen fish - with traps or trawls - must be considered to see how each method affects lobster mortality. And because lobstermen have been protecting females for a half-century, the help of fishermen who can efficiently collect real-time data on ecological conditions must be respected and eagerly recruited. Finally, awareness of how land-based contaminants might affect the productivity of inshore nursery grounds must be expanded and promoted.

Developing partnerships to seek answers to our questions became the central purpose behind the NOAA-NESDIS Penobscot Bay project, an intensive feasibility study of how decision makers might manage a marine fishery using ecosystem principles and techniques.

Monitors and other equipment aboard ALICE SIEGMUND record images collected by an underwater side-scanning sonar unit.

PIECES OF THE PUZZLE

The Gulf of Maine's scientific community is large and experienced, but widely dispersed. In Maine, the Penobscot Bay project recruited an interdisciplinary team of two dozen scientists to create a "virtual" marine research institute, each member of which would be responsible for contributing data on pieces of the bay's marine ecosystem. Woven into the fabric of this investigation was the data being collected by a larger group of more than 75 fishermen whose boats cover all corners of the bay. The fishermen would contribute information on what they caught and in some cases returned to the bay.

The scientists are from the University of Maine's Darling Center, the Bigelow Laboratory, the state Bureau of Geology and the Department of Marine Resources. They also come from the Island Institute, Maine Maritime Academy and the Lobster Conservancy.

Oceanographers think of the waters of the bay in layers. There's the top of the water with its tides and temperatures, the input from the Penobscot River and its associated estuaries. There's the bottom of the bay, once dry land but now an inland sea, with its different substrates and bottom communities. There's everything in between, including the way circulating water distributes nutrients, larvae, pollutants and benefits throughout this mini-system.

WHAT THE SCIENCE IS SAYING

Because lobsters spend most of their lives crawling around on different substrates of the bay's bottom, and because underwater photography has given us actual pictures of the ecological communities here, the bottom is a good place to begin constructing a detailed picture of the bay.

August 22, 1998: Joe Kelly and his wife, Alice, from the University of Maine, and their colleague, Steve Dickson from Maine's Bureau of Geology, are aboard the ALICE SIEGMUND. This 36-foot lobsterboat has been converted to an efficient research vessel by her captain, Corrie Roberts of the Island Institute.

The team has spent about 50 days over two field seasons collecting side-scanning sonar images of the sea floor. They do so by towing a torpedo-like instrument behind the boat that "sees" ("images," oceanographers might say) wide swaths of the bay's bottom and records a picture based on reflected sound. (Hard ledge gives a very different sonic profile than soft mud.) Sonar can also distinguish sand from gravel and can locate boulder fields - important habitat for juvenile lobsters. And because the ALICE SIEGMUND is smaller than most research vessels collecting oceanographic information, she's more cost-effective to operate. Today Kelly, Roberts and their team will deploy an ROV (remotely operated vehicle) equipped with underwater video cameras to "ground truth" some of the coarser sonar data they have collected on earlier voyages.

Out at the edge of the Muscle Ridge Channel, the ALICE SIEGMUND throttles back while one of the technicians starts the generator. The hydraulic gallus mounted on the stern picks up the ROV, which looks just like the miniature yellow submarine it is. With one hand on the joystick and the other on the color TV monitor, the ROV operator "flies" it to the bottom. Soon, the picture on the TV monitor in the pilothouse is a lobster-eyed view of the bottom and its complex habitats.

Kelly has concentrated much of his effort in the 30- to 60-foot depth bands of the bay, which are of particular interest to lobster scientists because a lot of the inshore fishery occurs here.

After two seasons of field work, Kelly has directly imaged about 30 percent of the bay. He has found several features of particular interest, including extensive pockets of mixed boulders, cobble and gravel in a large area south of Vinalhaven that have never been mapped before - ideal habitat for the frequently immense populations of juveniles and adults that are found here, year after year. Kelly has also found extensive gravel beds southwest of Vinalhaven, as well as a deep valley running down west Penobscot Bay. The valley, which corresponds to historic spawning grounds of cod and haddock in that area, consists of very hard gravel bottom scoured by strong currents.

LOBSTER BIOLOGISTS

It is not surprising that Maine is blessed with a strong contingent of lobster biologists. The fishery, after all, is the state's largest and most valuable. The most visible and outspoken of these biologists is red-haired Bob Steneck, who teaches at the University of Maine's School of Marine Sciences. He conducts his research from the Darling Marine Center on the Damariscotta River. For more than a decade Steneck has fielded teams of divers to sample lobster populations *in situ*, underwater, by counting and measuring all lobsters in small sample areas. Together with colleague Rick Wahle, now at the

Side-scanning sonar being lowered into Penobscot Bay

Along most of the Maine coast, the greatest number of very small lobsters live in one relatively rare kind of habitat, subtidal cobble and boulder fields in fairly shallow water. Most such areas lie around the islands and along the rocky shorelines of midcoast Maine.

Bigelow Laboratory in Boothbay Harbor, Steneck has pioneered a suction-sampling technique to collect tiny lobsters and determine where they spend the first several years of their lives.

Wahle and Steneck have come to a startling conclusion: along most of the Maine coast, the greatest number of very small lobsters live in one relatively rare kind of habitat, subtidal cobble and boulder fields in fairly shallow water. Most such areas lie around the islands and along the rocky shorelines of midcoast Maine.

Wahle and his colleague Lew Incze, also of Bigelow Lab, have discovered another important pattern. Near Damariscove Island off Boothbay Harbor, they consistently found many more newborn lobsters off the east-facing shorelines than off corresponding west-facing shorelines a quarter mile away. To Wahle and Incze, this finding suggests that near shore ocean currents are the lobsters' larval delivery mechanism. Inzce has proposed an intensive larval sampling program coupled to a study of how surface currents interact with local sea breezes, the "smoky sou'westers" and onshore breezes of summer days, to influence the direction of surface currents where young larvae are floating.

Rounding out the lobster team is Diane Cowan, who founded the Lobster Conservancy a few years ago in Harpswell, Maine. She has been working at the edge of intertidal habitats, where no one has looked for young lobsters before. Cowan has found a remarkable abundance of lobsters, ranging from recent settlers to nearly mature lobsters. Working with trained volunteers in a handful of coves at the lowest lows, or "moon" tides, Cowan has captured and tagged thousands of lobsters to discern patterns. Cowan has built a picture remarkable in its novelty, a convincing hypothesis that the nursery grounds of larval lobster extend upwards from 30 feet to the edge of the normal low tide mark.

Now Maine's chief lobster biologist, Cowan believes it is possible to use the sons and daughters and wives of fishermen, as well as others, to generate a predictive index of juvenile lobster abundance from year to year.

FISHERMEN AS SCIENTISTS

Walter Day has fished from Carver's Harbor, Vinalhaven, for 35 years, ever since his father helped him set off in a skiff to trap lobsters on short warps. In recent years, Day has fished a large area to the west and south of Vinalhaven alongside roughly 175 others from the island whose territory extends out to the far edge of Penobscot Bay. Under Maine's new lobster zone management system, Walter Day has been elected from his island district to represent the views of fellow fishermen on the region's Lobster Zone Council, which is intensely concerned with how the lobster resource of Penobscot Bay is managed.

Day became interested in the Penobscot Bay project when he was approached about helping to assemble a picture of lobster distribution and population structure in the bay, based on fishermen's records. An engaging and cheerful man, Day points out that in all his years of lobster fishing, he had never before been asked to contribute data to a scientific project. He was eager to cooperate. Day worked with Carl Wilson, a graduate student of Bob Steneck's now employed by the Island Institute, and a team of interns aboard lobster boats to record not just what fishermen landed at the dock, but also the number of the larger and smaller lobsters and "V-notched" females that had been marked by a cut in their center tailfin at some point during the past three years. These females were all carrying eggs when they were originally caught and marked, and at least in Maine, must legally be returned to the water after getting a free lunch at the bait bag.

Day helped Wilson contact other fishermen on Vinalhaven who might want to participate. David Cousens, who is the president of the Maine Lobstermen's Association, and who fishes the Muscle Ridge Channel on the west side of the bay, helped contact fishermen in his area. Leroy Bridges from Deer Isle did the same thing on the east side of the bay, and soon scores of fishermen began to pool their data.

Carl Wilson (right) heads up a team of interns working as sea-samplers aboard Penobscot Bay's lobsterboats.

Interns have collected data aboard scores of lobsterboats in Penobscot Bay.

This remarkable display of industry support is indicative of the sea change that is underway between fishermen and scientists. Amazingly, this information has never before been collected on such a scale.

GETTING THE BIG PICTURE

The hub of the Penobscot Bay collaboration is the treasure trove of satellite imagery that NOAA has made accessible to the project. To begin with, Andy Thomas of the University of Maine established a link with NOAA's enormous AVHRR (Advance Very High Resolution Radiometer) data archive, showing sea surface temperatures at a scale of about a square kilometer per data point.

As part of the Pen Bay project, Thomas analyzed the NOAA archive of AVHRR images collected four times a day between 1988 and 1995 to understand the changes in the location of surface currents over time. Using approximately 8,000 of NOAA's images, Thomas derived seasonal synopses, clearly showing the variations in the distinctive cold water surface currents that drive the ecology of the inner edge of the Gulf of Maine.

Because AVHRR data is meant to sample ocean surface features on a large scale, finer-scale satellite information must be examined to see how currents trending past the outer edge of the bay interact with Penobscot Bay itself. At the Island Institute, Dierdre Byrne and Chris Brehme researched the archive of NASA's Landsat imagery, which has much finer resolution - approximately 10 times the detail of the AVHRR pictures - looking particularly for patterns of persistent temperature features that might be correlated to lobster movements and distribution. Although Landsat has better spatial resolution, its temporal resolution (the number of images collected by the sensors) is poorer. Landsat only revisits an area of the globe once every 17 days, and unlike AVHRR cannot "see" through clouds and fog. Over a decade, only about 40 Landsat images could be assembled into seasonal averages to show in finer detail where the Eastern Maine Coastal Current is "entrained" in Penobscot Bay. Nevertheless, the patterns are startling.

SEASONAL PATTERNS

Until now, no one has definitively ascertained whether the Penobscot River plume of fresh water flows into east or west Penobscot Bay or whether, indeed, the pattern of river flow varies from season to season. Stream measurements show that during the spring freshet (March-May), the river's flow is more than twice what it is during the remainder of the year. It peaks at over five times its average flow.

Data collected during hydrographic cruises by University of Maine oceanographer Neal Pettigrew and correlated with satellite image analysis begins to reveal important relationships. When river flow is high during the spring, cold freshwater rides over the denser saltwater and empties into the upper portions of both east and west Penobscot Bays, freshening the upper surface waters by as much as five parts per thousand. In west Penobscot Bay, this fresh layer, which is confined to the top 15 feet or so, is discernible as far south as Gilkey Point on Islesboro and Northport on the mainland.

Despite the clearly detectable plume of cold river water between Sears Island and Islesboro, satellite image analysis of spring sea-surface temperatures shows that a temperature "front" forms across both east and west Penobscot Bays, separating the warmer upper bay from the colder outer bay.

There is also evidence of a clockwise "gyre" around Vinalhaven and North Haven. This surprising and previously unknown feature appears to dominate the circulation in the outer bay and control the exchange between the Eastern Maine Coastal Current and Penobscot Bay.

During the summer, sea surface temperature images show only a weak outflow from the Penobscot River. Along the western shore of the entire west bay, from Belfast Bay all the way south to Owls Head, relatively warm water is present in most images.

A temperature front between North Haven and Islesboro is a persistent feature in every summer image. It separates warmer waters to the northwest (around Islesboro) from cooler waters in the mid-bay off North Haven and Vinalhaven. During the entire summer, the colder waters of the Eastern Maine Coastal Current (the portion of the Gulf of Maine Gyre that runs along the eastern Maine coast from Grand Manan as far west as the outer edge of Penobscot Bay) are persistently visible in sea-surface temperature images. This current flows past the outer shores of Isle au Haut, Matinicus and Monhegan, but also intrudes far up the western shore of Vinalhaven, all the way north to Crabtree Point off North Haven.

Early in the year, outer Penobscot Bay is warmer than the northern and inner reaches of Penobscot Bay. But when the bay warms up during the spring, there is a vigorous mixing of waters, undoubtedly augmented by tidal energy and bottom topography. Vertical mixing is significant because it stimulates intense biological activity. In summer satellite images for areas where cold waters in the bay reach the surface, patterns can be seen where cold, nutrient-enriched bottom waters of the Eastern Maine Coastal Current are moving vertically to the surface. These areas are probably of importance to such species as lobsters, as well as juvenile herring, cod, haddock and other species.

In autumn, images demonstrate, the Eastern Maine Coastal Current turns offshore in a southerly direction at the eastern edge of Penobscot Bay, rather than beyond the western edge of the bay as is characteristic in earlier summer images. Nevertheless, colder waters are still found up against the shores of Vinalhaven and North Haven, and the characteristic front between the cold areas and the warmer, less saline waters of the western and northern portions of the bay is clearly present, though it is not as strong as in the summer season.

During the winter, a very distinct band of water flows along the southern or outer edge of Penobscot Bay. It is important to note that these waters are strongly differentiated from waters at the head of the bay, which are colder than the offshore waters at this season. During the winter, in other words, the estuary is the coldest part of the bay.

THE CIRCULATION MODEL

To round out the Pen Bay project, University of Maine Oceanographer Neal Pettigrew has deployed temperature buoys at the outer edge of Penobscot Bay and then in a number of locations within the bay to understand how water is moving throughout the system. These "smart" buoys collect temperature, salinity and current measurements at different depths in the water column on a continuous basis, storing them in an on-board computer, which then calls a University computer from a cellular telephone mounted on top of the buoy. Using this information, Pettigrew can calculate water mass movements.

Working with University colleague Huijie Xue, Pettigrew is constructing a dynamic circulation model, based on buoy data, reflecting the bathymetry or profile of the bay's bottom. The model includes data from wind sensors at Matinicus Rock and Owls Head airport.

Penobscot Bay, Pettigrew has found, does not act like a typical estuary because the mean flow is into west

Homarus americanus: the star of the show

Penobscot Bay, not out of it, as one might expect.

Pettigrew will launch additional buoys in the Bay during 1999 to get a clearer picture of circulation patterns, but his preliminary findings are immediately suggestive of a mechanism where larvae floating in the Eastern Maine Coastal Current get drawn into the western portions of Penobscot Bay.

For lobsters, which may be carried for more than 100 miles during their early life stages and then, responding to a temperature signal, drop out of the water column to prospect for hideouts, these current patterns suggest how and where things are likely to happen.

Pettigrew's findings also appear to correlate with Steneck's, Wilson's and fishermen's samples of subtidal lobster densities - highest at the outer edge of the bay near the border of the Eastern Maine Coastal Current, decreasing as one goes up the bay, lowest at the northern or inner edge of the bay. Lobster densities are also higher in western Penobscot Bay than in the east Bay, again suggesting that westerly flowing currents are driving their larval ecology.

A COMPREHENSIVE PICTURE

The Penobscot Bay project enters its third field season with a great deal of momentum. The project set out to deliver what may now be within our grasp: a comprehensive picture of how the complex oceanic dynamics interact with a regional embayment in the Gulf of Maine system. This picture might explain and perhaps predict the patterns of abundance for a commercially important species. This work is based on unique collaborations, not just within the scientific community, but in the fishing community, which needs this information to manage itself more effectively and predictably in the future.

Important as they are, lobsters are admittedly only one of the species that interact constantly with one another, and with the particular marine environments in which they are found, in Penobscot Bay. But by focusing on a single species, researchers believe they will uncover not just compelling new information on how the lobster moves through its environment over time, but assemble the many pieces of a puzzle in order to produce the ecologically-based predictive science that has always eluded fisheries management.

Philip W. Conkling is president of the Island Institute.

A Certain Obscurity

The experience of Swan's Island informs the poetry of Donald Junkins

CARL LITTLE

The Lobsterman Off Red Point
hauls his string of summer traps, and the buoys
line the seamless morning calm. Offshore
the whistler's ten second moan sounds another
story. Across the deck a small boy

learns addition and subtraction in Chinese. Later
he will make a drip castle on Fine Sand Beach
as the tide goes out. But now the morning's summer
eases into color and sound, teaching

the calculus of tone. Yellow jackets
prowl the railing spruce. Overhead a plane brackets
the blue sky looking for herring. You read in the sun,
your long black hair pony-tailed for fun.

- FROM *JOURNEY TO THE CORRIDA*, 1999

SWAN'S ISLAND is an island surrounded by islands. Once called Swan's Island Plantation, it is sometimes referred to as an archipelago or a "group" on account of its many satellite isles. Since the 1780s when the first bona fide settlers arrived, fishing and farming have been the principal occupations of island inhabitants, with some quarrying, boatbuilding, timber harvesting and, later, tourism complementing these livelihoods.

Historian Perry W. Westbrook was inspired to write the island's "biography" back in the 1950s. He was drawn to an isolated community that, in his words, "has retained so much of an earlier New England and America and which in its encystment in the sea exemplifies so well the adaptability of man to his natural environment."

Named for its Scottish-born owner, Colonel James Swan, in the 1790s, today the island is serviced by a car ferry that goes back and forth from Bass Harbor on the "back side" of Mount Desert Island daily throughout the year, with more trips in the summer. Over the last century boat service has been from a range of mainland towns, including Rockland, Brooklin and Stonington, a fact that underscores the island's somewhat unusual position straddling Blue Hill and Penobscot Bays (on a clear day, from atop Goose Pond Mountain, you can see both the Camden Hills and Cadillac Mountain).

My first crossing occurred this past July, a pilgrimage, you might well call it, to visit poet, professor and literary scholar Donald Junkins, who spends a part of each summer on the island. While assembling an anthology of Maine coast literature some years ago I had discovered Junkins' book *Crossing by Ferry* (1978) in a local library and had been stunned by the poetry, especially the verse consecrated to Swan's.

Arriving on the noon ferry and not finding my host at the dock, I began to walk. The islander who gave me directions to Junkins' home on the Red Point Road also warned me that it was a hike to the eastern side of Swan's. Had I consulted a map, I would have known that the island is quite large, with three "districts": Swan's Island proper, Minturn and Atlantic.

The poet and his son, Yunwei Chen, arrived shortly, I hopped in their car and we were off for an abbreviated tour of the island. We made two stops along the way: at the grocery store, where Junkins greeted just about every man, woman and kid he encountered, and at an island house, where he procured a container of freshly picked crabmeat, our lunch.

While personal friends with many Swan's islanders, Junkins enjoys a "certain obscurity" that allows him to interact with folks yet get work done. Indeed, he edited a groundbreaking anthology, *The Contemporary World Poets*, in 1973-1974, the one time he lived year-round on Swan's.

We were in the middle of a conversation on the deck of Junkins' house, the day I visited, when 80-year-old James Gillespie, a confirmed ex-urbanite according to Junkins, came walking out of the woods with a basketful of island-grown lettuce and Swiss chard. I had just asked Junkins how much time he spent on the island each summer, and Gillespie answered for him: "Not enough."

Junkins found his way to Swan's Island quite by accident. He was on a trip Downeast with friend and fellow poet Robert Bagg, who was looking for inexpensive land on Frenchman Bay. Through an Ellsworth real estate agency they heard about property on Swan's Island. They purchased a parcel together; Junkins eventually bought out Bagg's share.

There were numerous piles of slash when the poet first laid claim to the land, with three summers of burning required to rid the refuse left by pulping crews. At the same time, a brook was diverted to solve spring flooding problems.

The view from the house, which sits on ledge close to the water, is bracketed by Red Point and Black Point, and looks out on the Sister Islands. Junkins has seen deer swimming from the Sisters and once saw a whale pass by, "drooling fish," he recalls, as it rose straight up to feed. In the 1980s, he and his son Theodore set traps off the rocks, hauling in as many as 200 lobsters from the shore.

A sailor, Junkins kept a Muscongus Bay sloop in Russell Burn's boatyard for 25 years. He recently donated the boat to the Swan's Island Historical Maritime Museum.

Now in his late 60s but with the energy of a man 10 years his junior, Junkins built the island house himself, everything except the basic frame, the roof and the fireplace constructed of beach rocks. It's a simple and cozy dwelling, light-filled; the bathroom features the ubiquitous nautical charts for wallpaper.

Junkins does all the maintenance except the seasonal openings and closings, which he recently delegated to an island contractor and friend, Lonnie Smith, an ex-U.S. Marine whose wife and family run an island restaurant and bakery. This past summer he was replacing the wraparound deck. Bright new lengths of lumber shone among the gray of their weathered neighbors.

In some ways Junkins fits the type of turn-of-the-century Swan's Island vacationer as described by Charles McLane in *Islands of the Mid-Maine Coast*.

"As often as not, the visitor [to Swan's] would be not some rising industrialist eager to show off his yacht and his scrubbed children," writes McLane, "but an impecunious college professor who grew and sold a modest crop of berries or flowers to supplement his income." Westbrook confirms this tradition of professorial rusticators, noting that "many of the cottagers [on Swan's] are teachers."

Donald Junkins grew up in Lynnhurst, in Saugus, Massachusetts, not far from the Atlantic Ocean, but his Maine roots reach deep. Eight generations of his ancestors lived in York on the southern coast; indeed, the hearth bricks of the Junkins Garrison House (1657) are still in the ground on Scotland Parish Road.

The verse in *The Agamenticus Poems* (1984) draws on the rich history of this part of Maine. The poems have the flavor of John Berryman's "Homage to Mistress Bradstreet" or Robert Lowell's invocations of his New England ancestors. In fact, Junkins was Lowell's "star pupil" at Boston University in the 1950s.

Junkins's personal connection to the state began as a child. He spent weekends and summers with his family in Acton, Maine, on Great East Lake near the New Hampshire border. He still maintains a camp and tries to spend every August there. The collection *The Uncle Harry Poems and Other Maine Reminiscences* (1977), evokes this special childhood place, "the world of my senses," he has written, "uncomplicated without girls of my age there."

Junkins earned a bachelor's degree at the University of Massachusetts at Amherst and his Ph.D. from Boston University. He taught at BU, Emerson College and the California University at Chico before beginning a long-term professorship at UMass Amherst. Today, he is Professor Emeritus at his alma mater, where he also ran the MFA program in English for 10 years.

Scholars of American literature know Junkins from his significant work on the writings of Edward Taylor, the poet, and Ernest Hemingway. In the 1980s, Junkins served as guest

The poet and his son, Yunwei Chen. Junkins's wife, Kaimei Zheng, is from Beijing, China.

Deep June Onshore, Opening Up

The wind rushes the poplars, and the wild rose
ruffles by the rotting deck. I cannot see
Ram Island for the larch arms now between me
and old news. Whose death tolls

off Sunken Money Ledge? I have cut
a thousand branches to keep the lower
view. Earlier, two kayaks and their paddlers
gleamed from Red Point to the Sister Islands' hut.

Here my deck chair frays. The railing-mast
from the Muscongus Bay sloop floats
between deck posts, punky: my daughter's seven coats
of yellowed varnish peels. The past

is something else again. When I was a child
I found a Collier's magazine in a gold mine camp
above Lynn Bay, Alaska, Eddie Cantor
ogling on the cover, asking calmly wild-eyed,

"Will America ever be rich again?" Last night
at dusk I untied the nylon knot on the bleached-pole
nail and raised the Big Dipper toward the black hole
in the sky. Now in the morning light

the blue flag flutters against the spruce dark green
toward the open sea. Off in the woods a mourning dove
knits and pearls. I cannot measure the silence of this cove
as the noon tide comes in. Overhead a single gull careens.

- FROM *JOURNEY TO THE CORRIDA*, 1999

Junkins is little known in the state of Maine, even though last summer was his 30th on Swan's Island.

"Approaches to Blue Hill Bay": Chart No. 13313

Late June, walking the deer runs
to Goose Pond after supper
summer begins. Sidestepping
stormblown poplars,
dry-wading the slash from the pulpers' camps
ten years ago, keeping the imaginary
straight line from Duck Island Light to the north side
of Goose Pond Mountain in our minds' eyes; poking
our fish poles through young hackmatack
straight-arms, trying to keep from snagging
the green fur, the purple stars in the schooldesk landscape
of the nautical chart.

Yellow, blue.
The island woods are yellow. The evening sun
sprays through from the other side of the evergreens.
Woodcolors, our first grade pegs
arranging. We push for the first view
of the marsh-edged shore, spruce stumpsticks
edging deep water trout
neverminding the cold. We know where we are:
a mile straight in on the yellow.
We lose our way. My son climbs a blue spruce
to see where we've been: the two Sisters,
Long Island Plantation. On the left, the Baptist
church in Atlantic. We head into the sun.

Late June, walking the deer runs
to Goose Pond after supper
summer begins suddenly. We can hear
the creeing of gulls. Beyond the trees
they are landing, taking off, landing.
Saltwhite. Freshblue. It is all
prearranged. In a minute now
we will see the pond. Nothing has changed.

— FROM *CROSSING BY FERRY*, 1978

lecturer at the University of Freiburg, West Germany, with Hemingway the chief topic of his talks. The spring of 1993 found him at Xiamen University in the People's Republic of China as a Fulbright Professor. His second wife, Kaimei Zheng, is from Beijing.

The writer's curriculum vitae lists numerous poet-in-residencies, including stints at Bowdoin, Bates and the University of Maine's campuses at Gorham and Orono, as well as Harvard, Williams, Wellesley, Mt. Holyoke and such far-flung places as the University of Perugia in Italy, Fudan University in Shanghai and the University of L'Viv, Ukraine. Along the way there have been a number of distinguished literary prizes and a pair of NEA grants.

Junkins has published over 10 books of poems. His verse has been anthologized and has appeared in *The New Yorker*, *The Massachusetts Review*, *The Atlantic*, *Poetry* and *The Beloit Poetry Journal*. He is also a translator: his rendering of Euripides' verse drama, *Andromache*, was published by the University of Pennsylvania Press in 1997, and he has teamed up with his wife to translate the Tang Dynasty poet Li Po. His own work has been translated into French, most recently by Catherine Aldington, daughter of the British poet Richard Aldington.

On several occasions, Maine Public Radio has broadcast Junkins reading his poems. He also participated in several school visitation programs for the Maine Arts Commission, speaking and reciting his verse at high schools from Portland to Fort Kent.

Junkins is little known in the state of Maine, even though last summer was his 30th on Swan's Island. In recent years, he and his work have not been written up, nor have his poems appeared in the anthologies of Maine writings, an oversight in this writer's opinion. On the other hand, he hasn't gone out of his way to promote himself here; over the years, Swan's Island has served as something of a refuge.

Four of Junkins' collections feature island-inspired verse: *Crossing by Ferry: Poems New and Selected* (1978), *And Sandpipers She Said* (1970), *Playing For Keeps* (1991) and *Journey to the Corrida*, due out this year from Lynx Press. Another collection, *Late at Night in the Rowboat*, is in preparation for publication next year.

Junkins is a formidable prosodist, at times virtuoso, although he is subtle in his manner. His rhymes do not announce themselves; often only reading a poem aloud will reveal its inherent musical scheme. Proof of his poetic fearlessness are two sestinas in *Crossing by Ferry*, the title poem and "Markings." This complex verse form, first practiced by the troubadours, is composed of six stanzas of six lines each, followed by a three-line summary, or envoy. Rhyme is represented by a fixed pattern of the same six end-words in each stanza, but in a constantly shifting yet predetermined order. The three-line envoy contains all six end-words.

As in the best sestinas, the repetitions in Junkins' verse do not knock the reader over the head. The lines flow, unforced, although they fit a template. Here are the concluding stanzas of "Markings":

I watch the rockweed weaving in the tide.
A southwest breeze blows up, two gulls
take off. The gong off Sunken Money Ledge
sounds its iron sound, unlike my children
edging down the shore: they jump from rock
to ledge to rock, calling. Their voices

sound across the beach like porpoise voices
sounding in the blue. They stare at the tide-
line moving over the darkening rocks.
They keep track of their day: gulls,
boats, herons, seals — the way that children
do. They wave, crossing behind my ledge.

Summer children mark their island ways. The tide
marks every rock along the beach. Soon the nearby gulls
will watch from this ledge: I hear them now above my voice.

Junkins is also an inventor, creating patterns to fit his visions of the island. The configuration of the poem "Westward of Swan's Island," for example, literally echoes the shape of the coastline, points of land broken up by coves. Here are the final lines:

...We hoisted sails before
the beach.
The reach
of land called Devil's Head passed by. The shore

of one island became the shore of the last:
Scrag, Ringtown,
Gooseberry, John.
We ran before the wind as the sea passed.

Several poems draw on family experiences. "Swan's Island, Late Summer" was inspired by a baking business Junkins's daughter set up in the house (the profits from three summers' labor, the father recounts with pride, went toward her education at Mt. Holyoke). A more recent piece, "Lines Begun Near the Shore in a High Wind," is a moving portrait of the poet's mother that blends the elements of a summer island day with the musings of a woman drifting into what appears to be Alzheimer's.

Immediacy of experience marks these poems. In several recent island verses, the title serves as the first line, thrusting the reader into the setting and action of the poem. They capture the tangible impressions of the poet's island milieu, and the elements: "kelp flowing silky brown" ("Hauling Traps with Theodore: A Midnight Narrative at Low Tide"); "the scaffold silhouette of the buoy/bonging, moping over/the tide" ("Steamed Clams on Pond Island, the Quiet Drifting After"); and wild island strawberries ("Tasting Summer Fruit, Swan's Island"). They are accented with place names — Goose Pond, Duck Island Light, the Sisters, Sunken Money Ledge — which underscores the sense of place.

The poet has written that he considers himself a New England writer "by condition, not choice." He has lived most of his life in the region, and claims that during three years spent in California and one in West Germany he wrote more poems set in New England than he had during the rest of his life (one recalls that Ruth Moore wrote her first Maine novel, *The Weir*, while living on the West Coast).

Junkins is partial to the New England landscape, which includes the back coves, inner wilderness and tidal zones of Swan's Island. Yet he displays equal grace writing about other parts of the world special to him: rural Massachusetts, Key West, China, Bimini. His latest book, *Journey to the Corrida*, includes a remarkable group of poems inspired by the running of the bulls in Spain.

"What do they [the teachers] bring back to their classrooms, to their communities, and to their own daily living from these island summers other than rested minds and bodies?" Westbrook asks in his biography of Swan's. He answers his own question:

They bring a little of the repose of the spruce-covered hills and shores, the gladness of the bay at noon when the gulls dip and circle above the white-crested waves, the splendor of the water aflame at evening as the sun drops behind the western headlands, and the world-obliterating solace of the fog in late afternoon rolling in over the islands.

Surely the island-nurtured Donald Junkins would agree with these observations, although he might condense Westbrook's sentiments into a single word for what he carries away from Swan's: poetry.

Carl Little *is the author most recently of* The Watercolors of John Singer Sargent *(University of California Press). He directs the public affairs office and the Ethel H. Blum Gallery at College of the Atlantic. All photographs courtesy of Donald Junkins*

Lines Begun Near the Shore in a High Wind

The low roof pot swings
empty over the dark deck; two blue
flags snap-flutter on grommet rings
down the driftwood mast. "Who

are you dear?" my mother asked
last summer, eating her apple, her wig
awry under the sweeping gull's eye, basking
sweater-cozy in the sun. I was rigging

the extension ladder to stain the topmost
cedar shakes. "You mean I bore
you?" - and struck the driftwood railing post
laughing, nibbling her apple core

to the seeds. She dozed in the August
heat. I dabbed and brushed the dark
oil in as the breeze wavered in cooling gusts
off the water. In her sleep she mumbled "Hark,

hark." Later she said, "You mustn't mind me, dear,
but where have I been all these years?"
Now the island weather has changed again.
The blue flags drape, and the fog is rolling in.

- FROM *JOURNEY TO THE CORRIDA*, 1999

The Mid-Summer Crows Pause on the Rockweed

and shove off, their raucous
morning caucus
now above deep spruce,
loose

as leads on a thrown net:
now set
in the dark place. Moving, they resume
their off-tune

talk onshore. In the rockweed, a mink
slinks
the morning further on, and disappears.
Here

and there a cormorant gawks and dives, his black
shadow track
now gone in the incoming tide, glistening.
I am listening.

- FROM *JOURNEY TO THE CORRIDA*, 1999

REMOVING TIME AND DISTANCE

As Maine's year-round islands launch themselves into cyberspace, there's much to gain and lose - and plenty of healthy uncertainty.

BRIAN WILLSON

Does geography matter?

NOT, THEY SAY, if you're talking about the Internet. E-mail, discussion lists, live chats, and that user-friendliest face of the 'Net, the World Wide Web - no one cares if you're in New York, New Delhi or New South Wales. In the last half of the last decade of the 20th century, the Internet has blown like a cyclone, propelled by winds of its own momentum, growing like crazy, changing the way the world communicates, learns, buys, sells and conducts day-to-day business. As long as you've got a computer and can get connected, the rest of the world is but a few seconds and mouse-clicks away. These days, in fact, all you need is battery power and satellite bandwidth, and you can access the 'Net from any camel in the desert or any ship at sea. No place is too remote or insular. All humankind can in a heartbeat be - theoretically, at least - interconnected.

Yes, cyberspace has even come to Maine's year-round islands - or vice versa. But will it prove a godsend to these naturally isolated communities, bringing breaking news and information, tapping heretofore unknown revenue streams, linking inhabitants to the words and pictures of distant family and friends? Or will folks who've inherited or opted for an out-of-the-way lifestyle resent the unwanted interconnections? Will the 'Net prove curse or blessing? Boon or bane?

From the comments of a few Maine islanders with Internet experience, the answer would seem to be: a little of both. But few will label such an increased nexus as fundamentally bad.

"Generally, I think it's been beneficial for everyone," says Mary Beth Dolan, librarian and member of Monhegan Island's Board of Assessors. Perhaps ten or 15 year-round Monhegan residents have computers now, Dolan estimates, and several of these are connected to the Internet. Although she has no computer herself, Dolan is one of many others who share the library's PC, acquired a couple years ago through a grant related to a telephone company rate-case settlement. Along with the PC came 'Net access, and - at least as long as electricity holds out - a dependable link with the rest of the digital world.

In summer, Dolan says, the library welcomes a steady stream of seasonal residents who wish to check their e-mail. And, like mainland libraries, Monhegan's plays host to the usual contingent of patrons doing on-line research. But it's in the dark off-season, Dolan finds, when the library doors are left open 24 hours a day, that access to cyberspace seems most important. She tells of one year-round islander who opposed the whole idea of an Internet connection initially but, come that first winter lull, found it a friend and assistant in genealogical research - and could often be found surfing the Web well into the wee hours of the morning.

"At some point your work comes to a standstill," says Dolan. "For people in the wintertime, it's really great, because it connects them to the real world - a different world."

"I love e-mail"

For most people, that connection chiefly means sending and receiving e-mail, a form of electronic communication that's far swifter than a "snailmail" letter but nowhere near as urgent or intrusive as a phone call. These days it's an acceptable - and sometimes preferable - way of contacting family, friends, business associates, clients, potential customers and others in our daily cybersphere. In contrast to the neighborly greeting of old, shouted across a street or lawn, physical proximity has no bearing whatever on e-mail, Internet chats, discussion lists or other on-line relationships; in fact, many longtime e-mail correspondents have never met in per-

Illustration by Robert Shetterly

son or even spoken on the phone. The cyber-connection by definition involves a "virtual" community, a common interest, a remote transaction, a telecommute. It also involves time. E-mail can prove notoriously addictive: obsessive users, this writer included, often find themselves having to slog through a hundred or more messages in a day.

"I love e-mail," says Dolan. She acknowledges, however, that many on Monhegan share the same e-mail address - the one at the library. Still, Monhegan residents are fortunate to have public access, however limited: they can also surf the Web at the library or the tiny island school. But for most Maine islanders, access to the Internet has proved a costly, iffy proposition, at best. On Swan's Island, in fact, a problem accessing the 'Net resulted in an aborted early attempt at e-commerce by one business.

John and Carolyn Grace, founders of Atlantic Blanket Co., were among the earliest anywhere to attempt what is currently one of the most promising business uses of cyberspace: e-commerce, or the actual sale of products or services on-line. Three years ago, John Grace recalls, his Swan's Island business responded to a solicitation to create a Web page advertising, for a monthly fee, Atlantic Blanket Company's high-quality product - fine blankets made by hand of wool raised in Maine.

"They designed what we thought was a wonderful page for us," says Grace. "But in more than six months on the Internet, we got two inquiries, one of them from a fellow in India who wanted to sell us silk in thousand-pound lots." Confronted by such a meager response, the Graces deemed the experiment unsuccessful. But theirs was not the only failed early attempt at putting the Web to work - far from it. And they remain open to the idea of Internet commerce. "We're very interested in trying to find an appropriate way to get back on the 'Net," Grace says. Perhaps most frustrating to them, he says, has been the difficulty and expense of just connecting to the Web in the first place.

Until recently, most Maine islanders had little choice: it took a long-distance phone call to connect via modem to proprietary on-line services like CompuServe or America On-line, or to any of several local Internet service providers (ISPs) on the mainland. And added to the phone charges were the fees of the service providers themselves - the cost of Internet bandwidth. Lately, new bandwidth options have emerged, from 800 number modem access to limited satellite connectivity to newfangled wireless links to the mainland. Over time, bandwidth can only improve, and it'll be the new, imaginative uses and applications that will define the meaning and importance of the 'Net to the islands. And these prospective, conceptual uses, already wide, are ever expanding.

A new community

One coastal service provider has some novel ideas about how the Web can serve and benefit Maine's islands and other communities. Ligature Inc. (<www.ligature.com>), a Camden company, specializes in the "community Internet station" - a concept that treats the 'Net as, in a term coined by company head Richard Anderson, a "fourth medium" that goes beyond print, radio and television. Ligature's first such station serves the town of Camden with "local, daily, useful news and information."

Anderson says his company is beginning discussions toward a plan that will serve Maine's islands together as one of three community stations on the coast, each with its own news reports, discussion groups and photographic tours. He believes such a service will bring island residents increasingly in touch with each other, as well as with businesses and services along the immediate coast.

"One of the key aspects of the Internet," notes Anderson, "is what I call the 'back to the future' phenomenon. That is, the Internet, a future technology, [will allow] us to go back to a level of individual attention, personal service, and interaction that we enjoyed decades ago." He envisions a sort of extension of the local community, with doctors making the on-line equivalent of house calls, pharmacists expanding prescription services and retailers offering "chore" shopping - complete with home delivery.

Mary Beth Dolan cites evidence of this kind of 'Net benefit already having an impact on Monhegan. Islanders with sudden or worrisome medical problems, she says, "have been able to allay their fears in some way by getting information" from health sources on the Web. She thinks prescription and other services could also prove beneficial - though she can understand why early attempts at providing Internet pharmacy services have generated considerable controversy. But even routine doctor's appointments can be difficult for islanders whose travel plans are at the mercy of the ferry schedule and the weather. As things now stand, Dolan says, islanders must often "wait for the next boat, get an appointment, then can't get back for a couple of days."

Beverly Johnson, mother of three and resident of Chebeague Island in Casco Bay, has wasted no time in helping connect her island community, via the Internet, to the rest of the world. In fact, Johnson's Chebeague Anchor Web page (http://web.nlis.net/~bjohnson/Chebeag.html) might be considered the equivalent of a Ligature-style site, but with a homespun flavor — offering regularly updated news, island links, photos, even stories and poetry, all written by Chebeaguers and/or about Chebeague.

Johnson describes her page is "a place people can see what's going on and keep in touch." She believes the Web can't help but prove increasingly popular among residents of — and visitors to — Maine's island communities. "It will help the year-round population by making them feel connected to the rest of the world," she notes, and help seasonal visitors by reminding them of favorite summers. "I get e-mail and letters from all over the world thanking me for keeping [visitors] connected to the island even when they can't be here." Johnson, whose children attend the island school, is also helping develop the school's Web pages.

Upsides, downsides

Ample evidence exists of the Internet's benefit to islanders. Island schools, like schools everywhere, have been early adopters of Internet technology. Dolan cites one recent episode that confirms both the educational risks and benefits of a "wired" island: one of the school's handful of students, taking an advanced math course through a Johns Hopkins program, lost his connection when the school computer went down; he brought his CD-ROM to the library PC, where the software managed to connect up easily, and his usual on-line lessons commenced without a hitch.

On the other hand, Dolan sees at least one downside in the marriage of island and Internet: "I think it can take a small community and expose it to the world [in a way that] is very intrusive." She cites, as an example, Monhegan Commons (<www.monhegan.com>), whose lively discussion board invites non-islanders to weigh in on island issues. "I loathe the chat room," she says. Nor is she crazy about the pictures posted on the site for the vicarious enjoyment of those from away. "If they want to experience Monhegan," Dolan argues, "they should earn it: drive for hours, get on the boat, throw up, lug bags - and then watch the sunset."

Peter Boehmer, the island resident who runs Monhegan Commons, concedes that the Internet could mean big changes to the island way of life. "For better and worse," he notes, "this technology will change islands even more than it will the rest of the world. Removing time and distance will do a number on island communities." In fact, Boehmer likens the 'Net to an island ferry: it makes offshore communities more accessible, less distant.

"Rural communities have the most to gain," Boehmer argues, "yet are slow to see what the [Web] and e-mail have to offer." He acknowledges, however, that the 'Net also threatens to "increase the rate of over access and exposure. Islands by definition are limited - the Internet by definition is limitless." And that, in Dolan's mind, is the greatest risk. "I just hate that part of it," she says.

As for the usual concern about possible access by young people to the Web's notorious pornography sites, Dolan says this hasn't proved much of a problem on Monhegan.

E-commerce

Beyond communication, education and research, perhaps the most profound and immediate potential effect on Maine's island communities is use of the Internet for business. In just the past year, the general public seems to have warmed to the idea of shopping on the Web in a big way, with secure credit card transactions for goods and services growing by a rate of 200 to 300 percent or more in just the past year, according to Internet statisticians. Businesses in Maine, with its off-the-beaten-path geography, have benefited markedly from this trend; islands stand to gain a lot precisely because of their remoteness. So long as shipping can happen dependably - and Atlantic Blanket Company's steady non-Internet shipments suggest that it can - fulfillment shouldn't stand in the way of a growth trend in Internet sales.

Shipments of perishable products, such as live lobsters, offer the greatest challenges, but businesses dealing in such specialty food items are among the most successful on the Web. While e-commerce sites such as Amazon.com (books) and CDNow (music) might monopolize the latest news reports, numerous other less ordinary products are getting snatched up in on-line transactions. Dolan herself, for instance, runs Wyeth Editions for artist Jamie Wyeth (www.jamiewyeth.com), an on-line gallery of prints, posters and signed reproductions that's proved profitable in just its first year.

Peter Boehmer believes the benefits of e-commerce will work both ways, as Web-savvy islanders take advantage of expanded on-line shopping channels as well as new marketing and sales opportunities. But he grants that the trend could widen the gap between technologically inclined and more traditional islanders. "The problem is, will islanders take to this medium or can they survive economically as e-commerce divides island economies?"

For better or worse, it all boils down to interactivity - the fundamental give-and-take, back-and-forth, buy-and-sell of the Internet. Other uses of the 'Net include Web polls, surveys and elections, in which users cast votes and otherwise register their opinions statistically; discussion lists, in which like-minded people from far-flung geographies share thoughts and opinions about common interests; chat rooms, in which a group of keyboard-toting 'Net users enter "live" comments between and among each other; Web cameras, whose lenses point to natural, indoor and sometimes X-rated scenes worldwide. Business leaders communicate among themselves and between each other; educators share knowledge; craftsmen trade tips and tricks. Gossips gossip. Businesspeople work from home. You'll soon find each of these 'Net scenarios, if you can't already, on Maine's islands as well as elsewhere.

Overall, island communities likely will have a lot to gain - and a little to lose - as a result of the on-line revolution.

"It's pretty amazing," Dolan concedes. "Generally I think it's been beneficial to people on the islands, but I'm still just sort of baffled by the whole thing."

Brian Willson (<willson@3ip.com>) is owner of Three Islands Press (3IP), a small professional Internet services company in Rockport (www.3ip.net). He formerly was managing editor of National Fisherman.

FRENCHBORO AT WAR, 1941-1945

The church bell rang every night

DEAN LUNT

THE UNITED STATES entered World War II on December 7, 1941, after Japan launched a surprise attack on Pearl Harbor, killing more than 2,300 people and destroying a large portion of this nation's fleet. The war ended nearly four years later in two stages. On May 7, 1945, Germany surrendered, ending the conflict in Europe. On August 14, 1945, Japan surrendered.

In all, more than 362,500 American soldiers died in battle during those four years. World War II was a galvanizing event for the nation and a generation. The United States emerged from the war a world power; the generation that emerged would epitomize the country for the next five decades.

World War II was also the first war from which I knew veterans. It was a war that remained etched into the American consciousness, even into my generation. Sixteen people with ties to Frenchboro, including some lifelong residents, served in the war. And for four years, islanders rang the church bell at 6 p.m. each night in their honor. Each island soldier also carried a prayer book given them by Gladys Muir and the Sunday school children.

Vincent Davis, c. 1942

Clarence Lunt and Vincent Davis, c. 1942.

The island veterans I remember best were my great-uncles. I knew them while I was growing up in the 1970s and while I was a teenager in the 1980s. By then, they were older men, or at least they seemed that way to a young boy. At the time, I rarely connected the world of war which I saw in history books and movies, with the world of lobstermen I saw every day. I simply considered these guys colorful old fishermen who taught me to swear, chomped on cigars, watched soap operas, told tall tales, observed life and sometimes worked.

The two men I knew best were Clarence Lunt, my grandfather's brother, and Vincent Davis, my grandmother's brother.

I also knew Erland "Manny" Dalzell, Vernon Dalzell, Lawrence Davis, Hugh Stanley, Thomas B. Lunt, Jr. and Malcolm Lunt. They were all relatives of some sort. Kenneth Gardiner Lunt and Ralph Stanley died before I was born.

But always, Clarence L. Lunt and Vincent A. Davis were there, fixtures of an early island childhood. For years it seemed that Vincent, Clarence (and non-veteran Cecil E. Lunt) were sentinels on the Lunt & Lunt dock - watching, always watching. It is only a slight exaggeration to say that seemingly every time I landed on the island at the family wharf, they were there, sitting on a bench that leaned against the building or on some old lobster crates, watching the harbor.

And most nights about dusk, you could count on them wandering down past our house toward Lookout Point to sit on the ledges or beside the old duck-shooting blind. It was long an island tradition.

My most enduring impression of Clarence, grandson of a Civil War soldier and great-great-great grandson of a Revolutionary War soldier, is this: black hip boots rolled down, a partly-chewed, half-smoked Phillies cigar, gray cotton baseball cap with a long black bill, green cotton pants and scruffy whiskers. He walked with a slow, slightly bent gait that was almost like a slow roll. He never seemed in a real hurry.

He and Vincent knew at least something about what I was doing. They knew about little league games, Mount Desert Island high school, American Legion baseball, Syracuse University, some of my girlfriends, etc.

My favorite baseball player in the 1970s was Boston Red Sox catcher Carlton Fisk. They told me almost daily that Bob Montgomery was better.

Clarence was not the typical island bachelor of the 1940s and 1950s who never strayed far from the island shores. He actually lived in Tremont and served as a selectman there. He also owned a fine fiberglass boat, helped run a seining operation and had two children.

The last time I remember seeing Clarence was in a parking lot in Bass Harbor overlooking the Maine State Ferry pier on a hot summer afternoon in 1988, the year I graduated from college. He was sitting in his car. I leaned through the window. We told a few jokes, had a short discussion about my plans. But as usual, nothing too deep or personal. Just daily banter about baseball, family, fishing, gossip, trivial events and nostalgia. It was our connection.

Clarence Lunt died June 13, 1989. He was 66.

incent Davis said he knew when I was on the island because he could hear a ball bouncing on pavement. He constantly rousted me from his apple trees: "Hey, get out of that tree."

When I helped run a takeout restaurant one summer he walked to the wharf every day after The Guiding Light signed off and ordered the same thing, three grilled hotdogs and a coke. He called me "Da-Dean-dos" and he called most young women "girl."

Vincent, a lifelong bachelor, was a legendary drinker, but he gave up alcohol sometime in the 1970s as his health grew worse. He was sober from the time I can remember or would know the difference.

Vincent had no running water in his small two-room cabin overlooking both the harbor and the crumbling Alec Davis homestead where he grew up and which now served as his fishhouse. He lugged water from my parents' house in gallon plastic milk jugs or in an uncovered red plastic pail that smelled like cigar smoke. He usually filled the jugs at our outside faucet, but when that didn't work for some reason, we filled them for him at the kitchen sink. I don't remember him ever coming much past the front door. He waited on the lawn or just inside the doorway.

When he grew older, we often filled the jugs ourselves and left them at the bottom of our long stairs. Sometimes we lugged them to his door stoop. I, too, never went past his front door.

Vincent was the last great island storyteller, with a keen memory and the unhurried time to talk. He knew more stories about Long Island and its people than anyone in the past half-century. I heard a handful of those stories - sometimes true, sometimes embellished - but I have forgotten nearly all of them.

Vincent Davis died February 20, 1996. He was 82.

THESE SAME MEN SAW THE HORROR OF WAR.

Vincent served as a staff sergeant - the highest rank obtained by an islander - on a cable-laying ship based in the Atlantic. He enlisted January 18, 1941, and left December 16, 1945. While he operated under constant threat of German submarine attack, he never saw any real combat and remained a war buff until his death. He talked about the heroes and the movies and the war.

But Clarence saw the war in all its breathtaking brutality. Clarence drove an amphibious assault truck in the South Pacific, helping carry troops, equipment and supplies as the U.S. Army swung through the South Pacific and into the Philippine Sea.

In May, 1942, the United States realized that using amphibious vehicles in ship-to-shore operations was crucial to success. The U.S. made plans for amphibious brigades, and the army combed the Atlantic Coast for fisherman, small-boat operators and yachtsman to operate its new land-water vehicles.

Clarence Lunt, 20 years old and a lifelong Long Island fisherman, was just the kind of person needed.

As a result, Clarence, inducted February 11, 1942, saw some of the war's most brutal fighting as the U.S. drove toward Japan, across the island swamps, dense jungles, coral atolls and Japanese guns.

As a member of Company F of the 594th Engineer Boat and Shore Regiment, Clarence fought through the Marianas Islands, the Bismarck Archipelago, New Britain, New Guinea and the Philippines. He saw the bloody beaches.

On one of the Marianas Islands in the tall grass, Clarence fought hand-to-hand. During a fight, Clarence was gashed by a bayonet down his side. Wounded and bleeding, he kept fighting and killed his Japanese opponent with his own bayonet. Clarence powdered and bandaged his wound, but apparently remained pinned in the tall grass by Japanese snipers for nearly three days until reinforcements came to the rescue.

He received a Purple Heart.

By the time the war ended, he also received an Asiatic Pacific Theater Campaign Ribbon with Bronze Service Arrowhead, an American Theater Campaign Ribbon, a Philippine Liberation Ribbon with one bronze service star, a good conduct medal and a Victory Medal.

Not that I heard him say much about any of this. In fact, I can't ever recall hearing a single word. Nor can many of his closest friends. "If people started talking about the war he would get up and leave," said John R. Lunt, Jr., his youngest brother.

He even refused to collect his war pension.

More than once, Clarence said, "I was there, that was enough."

Vincent Davis died February 20, 1996, in a Bar Harbor nursing home. His friend Clarence had been dead nearly seven years, his friend Cecil nearly four years. Vincent was confined to the nursing home for his final days. At times, near the end, he sat on the end of his bed while his mind imagined he was down at the wharf on Frenchboro, just sitting and talking and watching with his friends.

A native of Frenchboro, **Dean Lunt** *is a reporter for the Portland newspapers. The above essay is excerpted from his forthcoming book,* Hauling by Hand, the History of Frenchboro, Long Island *(Islandport Press). Books will be available by the summer of 1999 from Islandport Press, 22 Lunt Harbor, Frenchboro, ME 04635; also from the Frenchboro Historical Society and Lunt's Dockside Deli.*

Salt Water Windows

JANET REDFIELD

Photographs by Brian Vanden Brink

"Hello, Janet. I'm with the Maine Department of Transportation, and we've just selected you to create stained glass windows for the new Rockland ferry terminal and four smaller terminals." This was a terrific commission: part of Maine's Percent for Art program which funds art in public buildings. And the committee that chose my work, comprised of engineers, artists, museum and gallery directors, was an artist's dream come true. "We'll leave the design for the windows up to you," they told me.

Thinking about people traveling to and from Maine islands on the many ferries made me wonder whether these passengers were aware of the incredible diversity of life around them during their trips. Windows into the salt water was the first step: I showed travelers some of the fish that might be swimming under them. Then it seemed logical to feature lobsters, crabs, shellfish and some of the other sea life in Maine's coastal waters. A window with a greatly enlarged view of the microscopic phytoplankton that abounds in the bays would show the beauty and complexity of these tiny creatures. Ferry passengers usually see birds floating in the water, flying overhead and perched on docks, trees and shore, so windows of sea and shore birds were natural additions.

"Salt Water Windows" consists of six round windows, each 35 inches in diameter, located in the Rockland Ferry Terminal waiting room, with a five-foot six-inch diameter round window located over the main entrance. There is also a rectangular window in the North Haven terminal and a round window in Islesboro. Two windows each are planned for ferry buildings in Bass Harbor and Lincolnville Beach.

An artist working in stained glass, **Janet Redfield** *lives in Lincolnville.*

The waiting room at the Rockland Ferry Terminal

EVICTED

Continued from page 53

In 1906, a year after the state declared Malaga residents its "wards," the family of Captain George W. Lane of Massachusetts, missionaries who summered on Harbor Island, arrived on Malaga bearing bibles and books. The Lanes set up a school in McKenney's home and dispensed food and clothing, as well as sermons on thriftiness and morality, to the islanders. Three years later, the community erected a schoolhouse with locally donated materials and funds raised by the Lanes as well as island residents. The state paid for a full-time teacher. On June 3, 1911, a *Bath Independent* article lauded the school, noted that it had attracted a tuition-paying student from Phippsburg, and called the students "a bright lot."

"The Lanes' emphasis was clearly on the children," observes Mosher. "The women were taught how to sew and serve tea, and the men were apparently considered lost causes."

In addition to their charitable endeavors, however, the Lanes drew more publicity to Malaga.

"Children of nature"

Muckraker journalism of the early 1900s, born of political corruption and scandals, promoted an end-justifies-the-means style of reporting that often ignored facts and played upon people's fears and prejudices. Malaga and the state government, charged with the islanders' care, got tangled in its grip.

For instance, Lauris Percy's article, subtitled, in part, "The Home of Southern Negro Blood ... Incongruous Scenes on a Spot of Natural Beauty," depicted the Maine-born islanders as lazy, stuttering, southern-accented idiots who "would almost sell their souls for a cut [of tobacco]." The 1905 report was published in the weekly *Casco Bay Breeze,* owned by two men from Beverly, Mass. The newspaper promoted real estate and summer cottage development in Casco Bay.

William Johnson, a wounded Civil War veteran with a pension who served in the 54th Massachusetts Colored Regiment (featured in the film "Glory"), was the only islander known to have been born in the South. Johnson, a seaman, had settled in Harpswell long before the Civil War began and worked as a fisherman until the state took him to the Old Soldiers Home as part of the eviction plan.

John Eason, a skilled mason who also lived on Malaga, "spoke like an English butler — his diction was perfect, according to a woman who knew him," Barry points out. "There were problems in the community," he adds, but the islanders "weren't the kind of people the papers painted them to be."

Holman Day's 1909 article, published in *Harper's Monthly,* though more charitable than Percy's, proved more damaging. Day wrote, "They have married and inter-married until the trespass on consanguinity has produced its usual lamentable effects. They are as near to being children of nature as it is possible for people to be who are only a stone's throw from the mainland and civilization."

The possibility that inadequate prenatal care or a historic lack of education on Malaga (even though the state had been paying for teachers on other islands and in rural communities for some time) might account for the condition of some children — or Day's impression of them — wasn't addressed. Additionally, Day failed to note that marriage among cousins was not uncommon in rural communities throughout the U.S.

Eugenics

A few first-cousin marriages (legal today in Maine and other states) occurred on Malaga, according to the Darling family genealogy, which may be the basis of incest reports. But Barry believes the genealogy "is suspect. Why were people doing a genealogy on this family to begin with?" He suspects it was created and "used for eugenics reasons."

Offensive postcards of islanders, described by anthropology professor Nathan Hamilton at the University of Southern Maine as "classic eugenics images," were mass produced and widely distributed at the time. "The Deuce of Spades" depicts a child on a woman's lap. "The Tray of Spades" portrays a woman and two children behind a fence. What happened on Malaga exemplifies "pure, essential eugenics of the United States," Hamilton emphasized.

"Eugenics books blamed crime and poverty on people who had been inbred," Barry explains. "The idea was to remove decaying stock from society, to stop them from reproducing. It sunk into the medical establishment and social work for decades — until Hitler. We think of it now as a bad thing but, initially, it was used to try and discover why certain families were poor, why they weren't doing well."

Though islanders *never requested* government or charitable aid, Malaga's state-funded pauper account, including the agent's fees, soared from $339 in 1905 to $1,171 by 1910. The "relief" virtually crushed the island's formerly lean, but self-supporting economy.

State funds, however, benefited other locals. One physician, charging $4 per house call, billed the state $200. (A doctor in Phillips, in comparison, charged the state 75 cents per visit in that region of western Maine.) And one local store's receipts show orders for two mattresses and bedding sent to the same woman in a two-year period. As Day prophetically wrote in 1909, "Donations of money bring more harm to them than otherwise." Not all of the islanders accepted aid, but all would face exile.

Eviction

Within months of taking office in 1911, Gov. Plaisted, a former newspaperman, ordered a three-member committee of his Executive Council to investigate conditions on Malaga. He appointed Dr. Gustavus Kilgore, a physician and surgeon from Belfast, as chairman. Kilgore also chaired the governor's Committee on the Home for the Feeble Minded.

In June, after learning the state might not have the legal authority to forcibly remove residents, Plaisted instructed the Attorney General to determine Malaga's legitimate owner.

On July 11, 1911, Plaisted, Kilgore and several government officials visited Malaga. The governor praised the school, made suggestions for community improvements and led islanders to believe he would help to relocate them to another island if push came to shove, according to a report in the *Bath Independent.*

Within three weeks of the visit, the state determined the Perry family (ostensibly descendants of the man who received Harbor Island's deed from Isaac Darling in 1847), owned the island, and had the Perrys issue writs ordering Malaga residents to vacate by July 1, 1912. Mosher stated he could not find a deed linking Malaga's ownership to the Perrys.

The eviction orders were served by a Cumberland County deputy sheriff. Since Malaga lies in Sagadahoc County, many people over the years have speculated whether the legal action was, indeed, legal. But the peaceful islanders, politically powerless and, evidently, dealing with politicians more clever than they, must have been merely stunned. They did not protest the order; they never opposed authority.

In the records of the Executive Council, the island is not discussed in any detail until Nov. 24, when agent George Pease's report on "conditions" at Malaga was submitted for review.

Pease, of Popham Beach in Phippsburg, was a former blacksmith and selectman. A local store owner, he was appointed as state agent by Plaisted and served as chief advisor to Kilgore's three-

member Malaga committee.

In the report, Pease lists each household, details their relationships and — in the same paragraph — assesses their physical, mental and financial condition. Entries include a name or names and ages, followed by observations, such as "half-breed, well and strong," "octorooms [sic]," "cost the state $18.63 during the first nine months of 1911," "negress," "good to work," "white," "full-blooded negro," "[o]ne child blind, others bright and smart," "[n]early self-sustained, have cost state nothing but probably would sooner or later," etc.

Pease also diagnoses which people are "feeble minded," and recommends they be placed in the state's recently-built institution in New Gloucester. Earlier, the state had removed two children from a family that "had never received [financial] help" and sent them to the Home for the Feeble Minded, according to the document.

James McKenney and another island man "will not live long," Pease predicts. Both men are left off the list of who should go where (e.g., "Order Griffin family to Phippsburg"), and how much money the state should pay the family heads if they accepted a buy-out.

The other man Pease names as a potential fatality is the head of a household that includes a wife (45) and her older sister, one son (30), two daughters (28 and 18) and the allegedly illegitimate two-and-a-half-year-old son of the eldest daughter. None are described as "feeble minded," but Pease recommends the two daughters be sent to the home as "fit subjects." He recommends the baby be sent to a well-heeled woman in Portland.

(A 1911-1912 state report, incidentally, criticized the Home for the Feeble Minded for not admitting more women of childbearing age because, the document states, they were "the menaces" who bred the "imbeciles.")

Noting that state ownership could "prevent people from settling [on Malaga], and [allow the state to] turn off the undesirable ones," Pease submits "that the State could purchase the island from the Perry heirs."

Unanimously approved by Gov. Plaisted and the council, the report states that Dr. Kilgore and two council members will execute the order. Three weeks later, Pease removes the entire family of eight, including the father — who is alive, but weak from an illness — and the baby, from the island and escorts them to New Gloucester. Pineland records reveal another daughter, 14, and not registered on Pease's list, was delivered by him that day.

All but two family members died in the institution. The father died less than six weeks after his arrival, and his son, later diagnosed with facial skin cancer, died within months of the father. Rather than being sent on to Portland, as planned, "the baby," who was later considered a

Tombstones at the Pineland Center in New Gloucester bear the names of only those former Malaga residents whose corpses were easily identified, and the year they were moved to the mainland and re-buried: 1912.

"good worker" at Pineland's farm, died at 16 following a tonsillectomy.

After five and 14 years of incarceration, respectively, the mother and one daughter were released, "paroled" into the custody of a facility trustee.

"What happened to that family, to all those children who were taken to Pineland, is atrocious," reflects Barry, who examined the islanders' records when he worked at Pineland in the 1970s, and described them as "active and able-bodied."

On the December day Pease delivered the family to New Gloucester, the governor approved the purchase of the island for $471, including court costs, from the Perry family. Later, the governor appropriated $1,350 to pay off the remaining families on condition they vacate the island by July. He also ordered a $1,100 payment to the home.

In March, Nelson McKenny wrote to the *Bath Independent* about the trouble some islanders experienced in finding new homes.

"A few changes have been made on Malaga since I saw you last," *McKenney reports, sounding as if he is addressing the editor personally.* "Eliza Griffin has moved over to the main, but she visits the old home every day so not to be homesick nor to give up her rights.

"Deacon John Eason has given up preaching and prayer-meeting through the troubles on his mind of leaving his old home and seeing all the old folks go too. And William Griffin and George Marks, after a long search, found nobody would keep them but everyone wanted them to keep out of their way, so these two natives are going to bunk on Hermit Tripp's little island up the New Meadows River where their minds will be easy.

"Another dried up native says how he will be the last nigger that leaves his old home after seventy years, the most of it on Malaga. The others of us are having hard times to find homes anywhere; all on account of folks saying we've got the cramp-catch in our fingers and take too many things that are lying around loose. But it's all a lie; we don't steal if we are poor.

"Uncle Jim McKenny [sic] is taking down his house today and Professor Eason will go next. If you know any place where I can crawl in with my wife and five kids and my old peg-leg please let me know."

By July, the *Portland Evening Express* was cheerily informing readers about islanders settling ashore. "Robert Tripp," the paper reported, "has built a house on a scow and with his family will enjoy houseboat life, moving about when and where he pleases."

Tripp's family of six sailed up the New Meadows River, searching for a place to land and hoping to secure a lot for his house. Barry recalls that "Tripp was prevented from landing by some good Christians and town leaders, so he hawsered up to trees on Bush Island. His family came close to starving several times."

Pointing to a December, 1914, news story headlined "Maine's Misery as Dark as Belgium's," which details the tragedy and compares it to that of German-occupied Belgium, Barry continued, "His wife, Laura, got desperately ill, so Robert rowed three miles in one of the worst winter storms in years. When he got back to the island with the doctor, his wife was dead. She's buried in a potter's field."

Tombstones

The state sold the island in 1913 to Everard A. Wilson, who occupied an office across the hall from Dr. Kilgore in Belfast. His sealed bid of $1,650 was $145 above the next highest bidder. Most bids were below $550. News reports note he planned to erect a "clubhouse" on Malaga.

Only two relics of Malaga's community exist today: the red schoolhouse, which was dismantled and reconstructed on Louds Island in Muscongus Bay, and the small gray tombstones that rise from the back row of Pineland's cemetery, which mark the graves of the uprooted dead. Far from the island on which they lived, worked and played, where they raised families and built homes, their markers bear the names of only those whose corpses were easily identified and the month and year of re-burial: Nov. 1912.

Neither the islanders nor their descendants have sued the state or appealed to the legislature for reparations. None have been offered during the past 87 years.

Deborah DuBrule writes regularly for Island Institute publications.

EYE OF THE *RAVEN*

Continued from page 5

Because I had acquired the rudiments of navigational knowledge, first in the old days with dead reckoning and later with FISH HAWK's comparatively simple Raytheon radar, we left RAVEN's complex set of interfaced navigation electronics to the last, and then ran out of time before the first trip. This of course fell on a day of early June murk and fog, but RAVEN and I ran down to Allen Island and back, and we bonded, big time. We made only one mistake, she and I, late in the day on the return trip to Rockland, in the Muscle Ridge Channel. I will not take you through the channel's several tight buoy turns except to say there came a point when, unclear of our exact location at a critical instant, I should have switched from the electronic chart — with its blinking cursor showing our location plus or minus 100 feet — to the radar to get a better fix. Apparently this was the part of the tutorial that I hadn't gotten wired quite right in my own mind, because the radar was in "course up" mode while the electronic chart was in "north up." Thus in the murk, when I saw the spindle on the starboard side that I had been expecting to port, the sensation of rock crunching through thick fiberglass seemed close at hand — though it wasn't, because there was good water on either side. Still, I knew RAVEN was urgently whispering to me about getting into the new age of marine electronics if the two of us were not to come to early grief.

The summer revealed brighter colors in lengthening light until the end of August, when we planned a voyage east, into the land of the ever-rising easterly sun and obdurate fogs. We had completed most of the critical tasks for the Penobscot Bay lobster project work detailed elsewhere in this year's *Journal* and wanted to check in with our partners downeast who were cooperating in a project to map herring spawning grounds along the coast between Beals Island and West Quoddy Head.

This region has been called the Bold Coast, where we know, at least from the written record, some of the Gulf of Maine's greatest concentrations of herring have gathered over particular bottoms to lay thick mats of eggs that once hatched in their trillions. For several years, fishermen had been telling us that these schools, pockets and coves of herring had become increasingly scarce. Because herring are widely considered to be the ecological linchpin in the Gulf of Maine, these early warnings were deeply troubling.

Last year's fieldwork was coordinated by retired lobsterman Stillman Fitzhenry of Cutler, who recruited 18 other lobstermen to participate by simply tending their traps and calling him when they found eggs on their gear. This enabled Fitzhenry to get to the location of the gear with a handheld Global Positioning System (GPS) unit, and then to sample the area nearby to estimate the area and depth of the egg mat. Stillman's daughter would enter the data on a spreadsheet and then send it to us at the Institute's office in Rockland, so we could generate maps. At the end of the season a year ago, the maps revealed a significant contraction of spawning areas in eastern Maine.

We leave astern the Isles des Monts Deserts, as Samuel de Champlain originally named them, and lay a course for Schoodic, obscured in thin and wispy fog. On board is Bill MacDonald, the Institute's Marine Resource Director. David Platt, editor of *Working Waterfront* and director of the Institute's publications, would join us in Jonesport. Chris Brehme, the Institute's Geographic Information Systems (GIS) specialist, would meet us in Cutler at Stillman Fitzhenry's wharf, where Fitzhenry and his family buy and ship lobsters and where we could refuel. Two days later, after stops at Cape Split, Beals Island, Jonesport and Northwest Cove of Cross Island (where we cruise in ghostly fashion by the Atlantic Salmon pen site in a dungeon thickness), we make our way into Cutler harbor.

We see precious little shoreline east of Schoodic until we moor at the head of Cutler Harbor near Fitzhenry's lobster wharf. By this time, Hurricane Bonnie is forecast to skirt the edge of the Gulf of Maine and make a landfall on the Nova Scotia side. We do not want to be on Grand Manan riding out even an edge of a hurricane, and so we push the schedule ahead by a day and a half, telling Fitzhenry we will meet with him, if it's all right, on the return trip.

The next morning we plan to cross the Fundy Channel, but not before the captain is given special dispensation for an early morning run out to Western Head, the massive, black-jawed headland that shelters Cutler harbor. Now a protected fairyland forest of stunted spruce, British soldiers and draping old man's beard, Western Head was bought from developers by the Maine Coast Heritage Trust as the 1980s real estate bubble burst.

From the tip of the headland, despite the dense fog, I could sense Grand Manan's outboard shores lined up like an outsized version of eastern Maine's facing headlands. From here it is evident that for herring, cod, haddock, whales, eagles and seabirds, there is little that is recognizable as an international border dividing the bay.

We are not a half mile past the Hague Line when a Canadian Coast Guard vessel, the CAMILLA, sweeps along our starboard side to hail us and ask our business. As the captain steps out of the pilothouse with two armed policemen flanking him on either side, we are vividly reminded of the deteriorating times in the fisheries, which tend to make borders tense and the sharing of information strained. We explain our purpose and show them the only fishing gear we have aboard — a herring egg collector, designed by David Stevenson and his colleagues at Maine's Department of Marine Resources — and the CAMILLA lets us on our way. It would have been hard to report to our partners in the project, the Maine Sardine Council, the Downeast Lobsterman's Association and the Maine Lobsterman's Association, that we had been turned away at the Canadian border.

Passing Grand Manan's southerly cliffs, we round up into Seal Harbour, mooring inside the towering government wharf that is in the process of being rebuilt. There we meet our Canadian partners, Janice Harvey and her husband, David Coon, of the Conservation Council of New Brunswick. Janice grew up on Grand Manan where her father, Mancil Harvey, once ran the local herring plant for Connors Brothers. Janice and David have arranged for a meeting room at a boatyard in Grand Harbour, but they are not especially optimistic that we will get a receptive hearing on the herring project. Fishermen on Grand Manan have hard feelings about environmentalists, and are especially wary of an outside group of whale researchers who have set up an office on Grand Manan. This office, they fear, is aimed at shutting down the herring fishery that has occasionally entangled whales in its nets.

We go to Grand Harbour at the appointed hour, but our spirits sink at the sight of an empty parking lot — not a single car or pickup in view. Because the boatyard secretary has forgotten to leave the key to the meeting room, two of us go to phone, while Bill MacDonald and David Coon wait in the parking lot hoping someone will show up. When Janice and I return, a knot of six herring weir tenders is in the narrow corridor with Bill and David, and a good deal of close-range measuring-the-cut-of-the-jib has begun. We sit down in the meeting room and discuss our methods and results for several hours. We answer questions. By the end of the evening, we have at least succeeded in introducing ourselves, even if we hadn't a clue as to the reaction to our admittedly strange-seeming group from across the Hague Line.

By this time, as we feared, Hurricane Bonnie is on everyone's mind. We plan to depart the next morning, anxious to re-cross the Fundy Channel and find a safe moorage. Even so, a nasty easterly wind is beginning to build, so we run up under the lee of Grand Manan's massive basalt cliffs, 12 miles along a stunningly bountiful and beautiful shoreline. Three minke whales roll up in our wake, while murres and a shearwater wing by to port and starboard. We spend the next day and night in Eastport riding out the gale and lashing rains, while Bill MacDonald visits with weirmen along the Perry shoreline at the entrance to the St. Croix estuary, where herring had once been netted by the millions, but have not appeared in the past half decade.

As we depart Eastport, we slide past the new cargo pier at Shackford Head and put into Lubec's impressive new town marina, where David Platt delivers the latest edition of *Working Waterfront*, something he has done at each port of call. The near-instant recognition of the paper everywhere we've been, including Grand Manan where we've just begun to distribute *Working Waterfront*, suggests that something is working for us.

From the marina we walk back along the waterfront to get a tour of the last remaining sardine packing plant in the area, Lubec Packing. Plant manager Peter Boyce gives us a tour, as well as a look at Hank Stence's remarkable experiments with spawning urchins.

Confident that the worst of the weather is behind us, we run RAVEN back out into the fog and murk of Lubec Channel, comforted by a flooding tide for the long run back along Fundy's starboard flank to Cutler Harbor. As the West Quoddy Head Light was about to disappear into the mist, I looked aft and saw, framed in RAVEN's rolling cabin doorway, Bill MacDonald with an open can of sardines from Lubec, taking his breakfast in seamanly style. The view is comforting; I would soon be turning over the helm to Bill's undoubtedly capable, baity hands.

With the forbearance of the crew, we set a course for Boot Cove where, it being Sunday morning, I want to observe a special moment of thanks and where Bill MacDonald wants to check in with David Pressley, the owner of the cove's large weir. This little left hook of a cove isn't much to look at on the chart, and I've never put in there before, so I don't know whether there'll be a lee from the big sea still running outside.

So we throttle back under the loom of the cove, listening to the surge of sea on Boot Head, straining to see a course, mindful of the tragically empty herring weir still faithfully attended by its keeper. We ghost through time and space, as if looking through a glass darkly, hoping for a quiet spot inside. There is the faint smudge of a shoreline, then a headland and finally the shrouded weir. We cut RAVEN's engine and lie there a-hull as the faint music of moving cobblestones rolls up and down the steep-faced beach. We stare down through the waters under the headland, watching the rhythmic pulse of the tides on the eerily quiet bottom of the cove. Boot Head has just been purchased by the Maine Coast Heritage Trust as a memorial to Peggy Rockefeller, who originally inspired a vision of protecting the long lengths of the Bold Coast after decades of cruising here in JACK TAR to find refuge and respite under its spiky headlands. We give our thanks for seeing this place, protected as it ever was and forever will be, world without end, amen.

We find our way back to Cutler Harbor where Chris Brehme, the Institute's GIS whiz, meets with us, Stillman Fitzhenry and Scott Sortman, a displaced but now local computer magician. They want to work out the details of a seamless transmission of herring-GPS data to Rockland to be translated by the wonders of Arc View software. This will go back to Cutler in map format, where Scott could then print a map Stillman will share with his corps of participating fishermen.

The big picture here is of a herring spawning cycle that once covered a much larger number of active spawning areas, mostly in less than 60 feet of water. In recent memory, spawning had stretched all the way west for another 60 miles to Petit Manan, began in late August and lasted through late September. Last year, the first spawning of the herring didn't come until October 9, and it ended on October 22.

It's hard to know what this means. Scientists might say it's impossible to know on the basis of two years of scant data. It's likely the herring are resorting to deeper water, either as an adaptation to avoid the purse seines that are mostly active in shoal waters, or to find cooler, deeper waters that are their signal to trigger the release of eggs. Either way, the system is changing in "real time" and we had better understand it if we're not to watch it disappear.

When we arrive back in Rockland, we find that Marge Kilkelly has been making headway on yet another Internet project. A few years back, during the heady early days of telephone deregulation, TDS, a telephone company based in upstate New York, bought the interchanges that encompass four Maine island communities, Matinicus, Isle au Haut, Frenchboro and Swan's Island. Needless to say, TDS made its purchase in spite of these four islands with their few hundred subscribers, rather than because of it. Phone service for islands has always been expensive and problematic, but islanders in these four most isolated places at the edge of the Gulf of Maine have been upset that their access to the Internet requires a long distance call.

At the Institute, Marge Kilkelly researched the range of technical solutions that might help these remote communities with their problem. She discovered (on an Internet site) that Mongolia had a similar problem, and that it had developed widespread Internet access across its remote geography, based on a wireless system. She then discovered that the local Internet service provider, the scrappy Midcoast.com, was interested in putting up a tower on Owl's Head. It didn't take too long before people began to realize that if there could be a wireless solution for midcoast Maine's most remote island communities, the same hardware could also provide phone service. And then a funny thing happened. The phone company decided that in interests of fairness, equity and keeping the government from doing something worse, it would agree to establish an 800 number for the islands to access the Internet without a long distance charge — *Mirabile dictu*.

Technology is always rearranging our lives, and whether it is for better or worse, of course, depends on your point of view. Cold War technology was repackaged and sold to the fishing industry, with disastrous results.

Yet we live in a region that also highly values continuity and tradition. There is widespread recognition that one of the keys to our future is to look and be different from so many other places in this once vast and diverse-seeming country.

At the Island Institute, we have invested heavily in environmental information technologies. We have worked to integrate them into an understanding of the changing world around us. We've developed large collaboratives — the waves of the future — and with our partners we've displayed pictures and lessons in schools, in new books and newspapers. We've helped create videos and held workshops. Now we are developing a new website, <www.islandinstitute.org>, that can pull in more information than ever, map it and then publish it more quickly and efficiently than ever before. We're not there yet; of course we'll never be there, because the wisdom of these new forms of information is that there is "no 'there' there."

Preparing for the future that is bearing down upon us, we're reorganizing ourselves. The old publications program will become a new information department, assuming responsibility for managing the design and editorial content of <www.islandinstitute.org>. To help pull in new information from community partners and orbiting sensors, we will hire an additional person with advanced training in remote sensing and geographic information systems. All our mapmaking and imaging capabilities will reside in the new information department. Community services will be developed into a community initiatives program which will, in addition to maintaining existing services, coordinate a new pilot community fellowship program where recent college graduates will work in communities and their schools, gathering and studying the resource information critical to understanding the changing environment.

Is the technology of the Internet, with its enormous capability of near instantaneous communications, driven by entrepreneurial American business models, going to preserve our diversity, as some claim? Or will it obliterate diversity, as others fear?

How we reconcile these competing forces is anyone's guess. But in the rapidly emerging future we are creating for ourselves, it's not a question of whether technology is problem or solution; it's both at once. We have to get connected just to find out; then, at an individual level, we become part of the problem. Or — just possibly — part of the solution.

REVIEWS

Acadia: Visions and Verse
Photographs and Poems by Jack Perkins
Camden, Maine: Down East Books, 0000
112 pages (hardcover), expected publication price $25.95

Reviewed by Carl Little

Jack Perkins is a veteran television journalist, having spent 40 years in the field, first as a correspondent for NBC and in more recent years as a host on the Arts & Entertainment network. He has a voice and visage of signature distinction — old school, like another Maine coast regular, Walter Cronkite.

Some years ago Perkins moved with his wife, Mary Jo, an artist, to a small island in Frenchman Bay. The place has worked its magic on the man, as a retreat from big city TV studios, but also as an entree, if you will, to the mighty charms of the Maine coast and more particularly Acadia National Park on nearby Mount Desert Island.

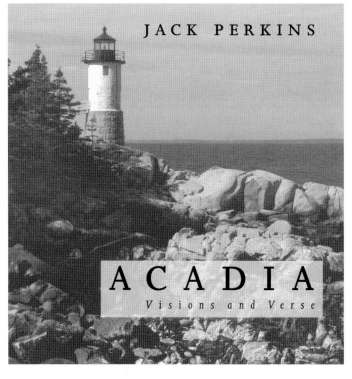

Perkins has been no hermit. He has proven himself a community man, willing to emcee a fund-raiser at College of the Atlantic or guest-narrate a concert at the Maine Center for the Arts in Orono. One of his favorite organizations, Friends of Acadia, fittingly garners a special plug at the end of this handsome coffee table book.

Mainers at large have come to know Perkins through a series of video portraits of Maine, produced by Dobbs Productions in Bar Harbor. His rich expressive voice perfectly suits dramatic aerial views of lighthouses, of Katahdin and Moosehead, of villages, towns and cities across our fair state.

In *Acadia: Visions and Verse,* Perkins adds two more hats to his collection: photographer and poet. He wears them with different degrees of authority and success, a fact underscored by his foreword to the book, much of which is devoted to an appreciation of photography.

In this field, Perkins has some interesting credentials. He once interviewed Ansel Adams, the great American master, and he attended the Maine Photographic Workshop in Rockport where he studied with several acclaimed practitioners, including Tillman Crane. His black-and-whites bear out this training: clear, sharp, well composed and of a fine silvery-gray tone. Like fellow Acadia fancier-photographers George De Wolfe and Dorothy Monnelly, Perkins has sought and captured the visual poetry of the park.

Perkins' subjects tend to be somewhat cliched: lighthouses, coastal scenics, a church interior, lobster buoys. The best pictures focus on natural patterns and elements — "hieroglyphs" etched in sand, angelic birch trees, the top of a mushroom, a knot in a tree. And there are some fine tributes to the park's famous stone bridges.

Photographs are legitimate prompts for poems: who hasn't admired the photo-inspired verse of James Russell Wiggins in the *Ellsworth American* each week? Yet such an exercise rarely leads to deeply felt poetry. On the other hand, Perkins, like Wiggins, is a deft wordsmith. Moreover, his lines are often imbued with the rhetoric of a man at awe with the beauty of his surroundings. He urges us to look, in line and in image, as in a spread devoted to asters in granite:

Rock speaking mutely of first things and last,
Of unforeseen future and unrecalled past;
Echoing Paleozoic disasters,
While presaging towers to be crafted by masters.

Blessed be those who, facing all this,
Somehow, wisely, all this can dismiss
               — And find asters.

Perkins is a conservationist; he cares about his surroundings (and his island home is solar-powered). This tribute to Acadia should gain the park new admirers and, one hopes, many new caring friends.

Occasional Papers: The Publications of the Islesford Historical Society

Reviewed by Carl Little

The Islesford Historical Society, based on Little Cranberry Island, is a model of its kind. Since its founding by Hugh Dwelley, Ted and Cara Spurling and Gail Grandgent in 1990, the society has grown and prospered, pursuing the history of the Cranberry Isles hither and yon. As society director Dwelley wrote in a recent letter, "We [founders] never anticipated 200-plus members, 16 publications and a History Room added onto the Islesford Library!"

The society's publications, many of them called *Occasional Papers*, have been printed with regularity since 1990. Each of these desktop productions sports a different colored cover (the latest is dark pink!) and is printed in what might be called "old-fashioned typewriter" font.

As Dwelley explains it, over the years the society has edited and published "whatever came to hand," including collections of poems (*Jabberwocky, George Gilley* and *The Medicine Chest*). They have dipped into architecture via Marion Spurling's *Houses of Islesford* and social studies, in John Brooks' *Rusticators at Large*.

Two recent publications reflect the thrust of the society's mission. *The Gilleys of Baker Island and Islesford, Maine*, by Dwelley, is one in a series of papers devoted to "Pioneer Settlers of the Cranberry Islands." The text is a mix of genealogy and history, with the occasional anecdote thrown in for good measure.

Most engaging are the accounts of island life and ways. For example, we read with fascination about the "wreck stripping" that went on with some regularity on the islands. Many a schooner ran aground on the coast, sometimes run onto the rocks on purpose, for the insurance.

Such episodes of cargo recovery are also recounted in *Cranberry Isles Sketches* by Louis G. Barrett, which is accompanied by a marvelous oral history given by Tud and Polly Bunker in Occasional Paper No. 9, published in 1997. Here one finds the tales that may have started out as fact, but that through telling and retelling tend to grow, shall we say, tall, à la Ruth Moore or Jones Tracy. As Dwelley puts it in a foreword, "While the stories and characterizations are all interesting and much is true, 'a grain of salt' is clearly in order now and then."

These texts cover the gamut of island existence, from the tragic death of young "Unkie" Alley in the harbor to a gripping account of a shark attack right out of "Jaws." Along the way we read about a dead whale, Tud Bunker's ingenious use of a broomstick to replace a broken water-pump shaft on a floundering craft and the quintessential Cranberry Island automobile:

Life Saving Station, Little Cranberry

The varnish was long gone [off Oscar Wedge's station wagon], the side curtains were all a-flappin' in the wind, no hood and a teakettle on the running board. Starting it was always a procedure demanding great patience of an onlooker since Oscar first had to get out a wrench and remove the starter motor.

Dwelley, who retired from the Foreign Service in 1988, devotes much of his time to these projects, including four years spent transcribing Vincent Bowditch's *Islesford Journals* (1894-1928). He has drawn on Ted Spurling's knowledge of the area (the latter has written a short history of the Cranberries) and researched the islands in libraries up and down the eastern seaboard.

Much of this research has been conducted to what might be called a greater goal: Dwelley's *A History of Islesford*, which is currently being edited for publication in 2000. Ted and Cara Spurling will be the final editors.

Publications can be ordered from the Islesford Historical Society, Islesford, ME 04646.

Carl Little *has written extensively on the art and literature of the Maine coast. He has contributed a foreword to a new edition of Rachel Field's* God's Pocket, *due out this year.*

ISLAND INSTITUTE
Sustaining Islands and Their Communities

The Island Institute is a non-profit organization that serves as a voice for the balanced future of the islands and waters of the Gulf of Maine. We are guided by an island ethic which recognizes the strength and fragility of Maine's island communities and the finite nature of Gulf of Maine ecosystems.

Along the Maine coast, the Island Institute seeks to
- support the islands' year round communities
- conserve Maine's island and marine biodiversity for future generations
- develop model solutions that balance the needs of the coast's cultural and natural communities
- provide opportunities for discussion over responsible use of finite resources, and provide information to assist competing interests in arriving at constructive solutions

The Island Institute is the sole organization focusing its programs and resources wholly on Maine's islands, their people and the waters that surround them. Programs are for year-round islanders, fishermen, students and teachers, scientists and resource managers, summer residents, island property owners, coastal communities, state and municipal agencies, boat owners and island visitors.

The Institute's schools and community service programs help island communities remain viable through:
- support of island schools
- scholarships for island students
- long-range planning and special economic development projects
- information resources linking island towns, property owners and state and federal resource agencies on a variety of issues, including transportation, water quality, and solid waste regulations
- legislative action

Programs in marine resources provide information regarding the challenges and opportunities facing fisheries, aquaculture and working waterfronts. The program staff:
- collects critical marine habitat data
- monitors marine conservation legislation
- helps manage a groundfish stock enhancement effort
- supports Gulf of Maine fishermen in researching and managing the fisheries upon which they depend
- tests new opportunities in small-scale aquaculture

Through its publications and information programs, the Institute produces the annual Island Journal and a monthly newspaper, Working Waterfront. The newspaper addresses issues which directly affect people who depend on the coast and marine environment for their livelihoods.

The Institute also publishes books on a regular basis. Significant titles include Penobscot: The Forest, River and Bay; Rim of the Gulf: Restoring Estuaries in the Gulf of Maine; From Cape Cod to the Bay of Fundy: An Environmental Atlas of the Gulf of Maine (with MIT Press); Killick Stones, a collection of island stories; and (with Tilbury House) Charles and Carol McLane's series on the islands of Maine.

The Institute maintains a comprehensive geographic information system database on the islands and waters of the Gulf of Maine, including natural resource data collected from satellite imagery. Visitors to the Institute's website at <www.islandinstitute.org> can learn more about the organization's activities and access a full online version of Working Waterfront.

The Maine Lights Program concluded successfully in 1998. Conceived by the Institute in 1994 and signed into law by President Clinton in October of 1986, this program grew out of concern for preserving an endangered and important part of the Maine coast's cultural history. Through the Maine Lights Program, the Institute helped place 28 Maine lighthouses into the hands of local non-profit organizations and municipal, state, and federal government agencies. By finding appropriate "caregivers," the program has ensured that these lighthouses remain accessible to the public.

Thirty-six percent of the Institute's FY 98-99 operating budget of approximately $2.5 million is expected to come from annual membership dues and from personal and corporate donations, 34 percent from foundations and 30 percent from earned income (publications, conferences, consultations, service contracts, etc.). The Institute's earned income is substantially greater this year because of a contractual agreement with the National Oceanic and Atmospheric Administration (NOAA) to provide marine-resource research services in Penobscot Bay. The Institute's annual report, listing members and presenting the financial picture in detail, is available upon request.

MEMBERSHIP

Membership participation from a variety of people is the only way to sustain a balanced organization, and we welcome your involvement in any capacity. Become a member — or call, write or stop in to ask for further details regarding our programs, or how you can help through donations or volunteering.

BOAT DONATIONS

Over the years, the occasional donation of vessels has significantly enhanced our programs. Such gifts result either in boats we keep and use, or boats we convert into the funds necessary to fulfill our mission. Either way, should you be in a position to consider such a gift, we'd like to hear from you.

ISLAND INSTITUTE
410 Main Street
Rockland, Maine USA 04841
phone (207) 594-9209
fax (207) 594-9314
email <institute@islandinstitute.org>
www.islandinstitute.org

ISLAND INSTITUTE MEMBERSHIP BENEFITS

❖ **ISLAND JOURNAL**
A nationally acclaimed annual publication, featuring island life, the arts, stories, photographs, essays, poetry, people, and marine science.

❖ **WORKING WATERFRONT/INTER-ISLAND NEWS**
Monthly news and information about islands and their communities and resources, marine industries and the coastal economy.

❖ **NOTES FROM THE FIELD**
Regular updates on Institute projects, research and activities.

ISLAND INSTITUTE • 410 MAIN STREET, ROCKLAND, ME 04841 • (207) 594-9209 • <www.islandinstitute.org>

ISLAND INSTITUTE PUBLICATIONS

1999 Island Journal, Volume 16
A Broadway producer takes the Fox Islands and their students by storm; Ashley Bryan's art and stories; the Penobscot Bay Project reveals a bay's secrets; three island siblings trek Baffin Island; John Fowles' universal thoughts concerning islands. **$14.95**

Islands in Time: *A Natural and Cultural History of the Islands of the Gulf of Maine* • With Down East Books
By Philip W. Conkling • As "wild" as Maine's islands appear today, virtually all are affected by centuries of human activity. First published in 1981, this classic has been updated, expanded and redesigned. For anyone interested in islands, coastal communities, their inhabitants and the resources that sustain them. **SC $19.95 • HC $27.95**

Rim of the Gulf: *Restoring Estuaries in the Gulf of Maine*
Edited by David Platt • Illustrated with color photographs, maps and satellite images, this book documents the sorry state of the estuaries that ring the Gulf of Maine. **$19.95**

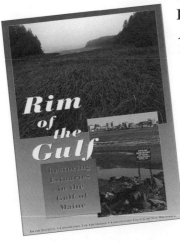

Penobscot: *The Forest, River & Bay*
Edited by David D. Platt • An in-depth historical profile of the region, profusely illustrated with photos, charts and satellite images. **SC $16.95**

From Cape Cod to the Bay of Fundy: *An Environmental Atlas of the Gulf of Maine*
Edited by Philip W. Conkling • MIT Press
SC $29.95 • HC $50.00

Sightings: *A Maine Coast Odyssey*
Peter Ralston's photographs of coastal Maine. Available signed by the author.
HC $50.00

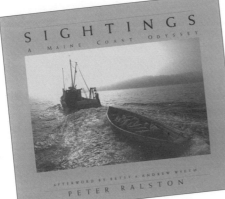

Working Waterfront/ Inter-Island News Subscriptions
One year (11 monthly issues): **$20.00**

Membership

- ☐ $40 Member
- ☐ $40 Non-profit organization
- ☐ $60 Subscriber
- ☐ $100 Contributor
- ☐ $250 Donor
- ☐ $500 Guarantor
- ☐ $1,000 Sustainer
- ☐ $2,500 Benefactor

Prices good through May, 2000. Canadian residents, please add $10 for shipping; all other countries add $15 U.S.

Name _____ Phone _____

Permanent Address _____

City, State, Zip _____

Summer mailing address* _____

City, State, Zip _____

Do you own island property? Yes _____ No _____ Island Name _____

Method of payment: Check _____ Visa/MC # _____ exp. _____

Signature _____

* Benefits will be mailed to this address from June 1 through September 1.

Island Institute is a recognized non-profit organization; however, under current IRS regulations, all but $34.95 will be tax deductible, reflecting the actual value of the membership benefits you will receive (see reverse).

ISLAND INSTITUTE • 410 MAIN ST., ROCKLAND, ME 04841 • (207) 594-9209 • FAX: (207) 594-9314

Publications Order Form

Quantity		Retail $	Total
____	ISLAND JOURNAL, VOLUME 16 – 1999	$14.95	_____
____	ISLAND JOURNAL, VOLUME 15 – 1998	$9.95	_____
____	ISLAND JOURNAL Back Issues • Volumes 2, 3, 5-11, 13, 14 available @ $7.95		_____
____	ISLANDS IN TIME	$27.95 HC / $19.95 SC	_____
____	RIM OF THE GULF	19.95 SC	_____
____	PENOBSCOT: THE FOREST, RIVER AND BAY	$16.95 SC	_____
____	FROM CAPE COD TO THE BAY OF FUNDY	$50.00 HC / $29.95 SC	_____
____	SIGHTINGS	$50.00	_____

Maine residents add 5.5% sales tax _____

Shipping ($4.50 for up to 2 items; $1.00 for each additional 3 items) _____

____ WORKING WATERFRONT/INTER-ISLAND NEWS • One-year subscription (11 monthly issues) $20.00 _____

TOTAL _____

Prices good through May, 2000.

Please ship to: Name _____

Address (please use street address for UPS delivery) _____

_____

Phone/Fax: _____

Method of Payment: ☐ Check enclosed ☐ Visa/MC # _____ Expiration date _____

ISLAND INSTITUTE • 410 MAIN ST., ROCKLAND, ME 04841 • (207) 594-9209 • FAX: (207) 594-9314